FOREST RESOURCE, ECONOMY AND ENVIRONMENT

FOREST RESOURCE, ECONOMY AND ENVIRONMENT

A.K.SINGH
V.K. KUMRA
J. SINGH

CONCEPT PUBLISHING COMPANY, NEW DELHI-110015

First Published 1987

Published by
Naurang Rai
Concept Publishing Company
H-13, Bali Nagar
NEW DELHI-110015 (India)

Printed in India
Offset Printing Centre
A-27, Naraina, Industrial Area Phase II,
NEW DELHI-110028

Foreward

Of the two important functions of forests, viz., productive and protective, the former has often been excessively emphasized because the direct economic benefits are easy to compute. In view of the accelerated depletion of the non-renewable resources, it is being increasingly realised that the hope for feeding and clothing the ever-expanding population of the world rests on the forest resources which are fortunately renewable. Further, the forest is the most crucial element in safe guarding the environment because of its role in providing long-term ecological security including conservation of our life-support system. Forest ecosystems are needed to protect wildlife and soil, to sustain man, to stabilize climate, to optimize water yield and to purify water. The long-term, indirect economic benefits of the protective function of the forests are staggering indeed. In order to perform the multiple functions which reflect the ecological imperatives, as outlined above, the forest has to be well-stocked and its cover must be adequate.

However, the short-term economic imperatives have led to indiscriminate, unscientific and over exploitation of forest resources. The delicate relationship between man and forests is shattered. Depletion of forest cover has accelerated landslides in the hills, and floods in the plains. Many of the erstwhile abundant species have become rare and several of them are on their way to extinction. In the long run the soil and gene erosion will also undermine the renewable base of agriculture.

It is unfortunate that many of us because of our limited understanding tend to view ecological and economic imperatives as antithetical. Like ecology the word "economics" is derived from the Greek root "oikos" (meaning 'household'), and the "nomics" means "management",. thus economics translates into "management of household" and ecology into "study of life at home". Evidently there should be no conflict between the ecological and economic imperatives especially where such a vital resource as forest is involved. A forest represents an investment of which the annual income (increment) cannot be separated from the capital (growing stock) which produces that income. Sound economic principles demand that the capital be preserved and a steady income be maintained. A sound forest management policy takes care of this and it views the "income" not only in context of pecuniary but also in context of environmental benefits, the latter being more important among the two.

I am, therefore, happy that the present book examines the role of forest resources in shaping both the economy and the environment of Uttar Pradesh. A balance in economic and ecological imperatives has been maintained throughout. A vast amount of very useful data has been collected and collated. Causes of deforestation and issues of forest management have been discussed in sufficient detail. The book is aimed at a wide audience including students and professionals in geography, economics and environmental science, and planners and decision makers. The authors, Drs. A.K. Singh, V.K. Kumra and J. Singh, are to be congratulated for bringing out this excellent and authoritative book.

Banaras Hindu University
May 2, 1986

J.S. Singh
Professor of Botany

Preface

Since the advent of civilization on this spaceship terra, forests have played a major role in maintaining the environmental quality and providing livelihood for human beings. During modern times the policies of rapid economic gains have wrecked the ecological balance through the process of over exploitation of natural vegetation. The exploding population and advanced technology have jointly brought the situation almost to the point of no return. Every day several species are becoming endangered or extinct. In fact, in the name of development we are shaking hands with destruction. It is being increasingly realised that if we are not able to save our environment, we will be the biggest loser. The environment, that feeds us, renders various materials, conditions and opportunities in which we live, breathe, eat and drink, is under tremendous pressure due to the speedy extraction of forest resources. If the present trend of deforestation continues, the hope for better future can not be more than a mirage. Thus, the problems of degrading forests and deteriorating environmental conditions require an urgent and serious attention for finding out ways and means to conserve the forest resources for our own benefits. To escape or to avoid these issues means to give an offer to so many problems in so many ways.

In the beginning, the plan of the book was to cover spatial and economic aspects of the forests only and the main issue of environment in relation to forest was not included in the original plan of the book. As the writing progressed it became evident that attention has to be paid to this aspect because otherwise it would have amounted to escapism.

This book is a realistic outcome of the longfelt gap of 'how geographers perceive forest as resource!' and 'what place forests occupy in geographical explanations?' Considering forest as one of the major environmental components, authors have tried to portray some major aspects, viz., spatial distribution, economics, depletion of forest resources and its consequences on environment and finally its management in regional perspective of Uttar Pradesh.

Uttar Pradesh, stretched over an area of 2,94,413 km^2, possesses a variety of natural landscapes and cultural cocoons, thus, reflecting an image of typically unique bio-geocenosis. A few decades back the State was full of natural vegetation and fauna but now the conditions have markedly changed, and that too for the worse.

Rapidly shrinking forest cover of Uttar Pradesh have induced various scholars and researchers to ponder over the issue. Presently, about 17.41 per cent of the total geographical area is under forest according to a government statistics. However, the satellite images present an entirely different picture (7.14%). The satellite images of U.P. Hills depict 28.7 per cent area under the forest while the government statistics show much higher percentage, i.e., 67.3 per cent. The data base of this book is mainly the government statistics but the same has been supplemented by additional studies and observations. The state fails to fulfil the national norms of minimum forest cover. Spatial unevenness and local/regional level shortage of forest resources vary at different levels and scales. The state does not present even 0.05 hectare forest land per person. Out of fifty-six districts, twelve districts have nominal or even no forests. Thus, in the districts where forests are not available the problems related to timber, fuel, fodder, soil erosion and flooding etc., are very acute.

This study intends to bring forth the systematic understanding about forest resources of the state, their role in shaping the economy and environment and lastly their management. The book has been divided into five chapters. The first chapter is devoted to important components of physical environment of the state which are responsible for the growth of vegetation. The chapter two focusses on the various aspects of spatial distribution of forests. The third chapter presents an insight into the various components of forest economics while the fourth chapter highlights the causes of deforestation and its ill-effects on environment. The last chapter encompasses issues of forest management in general rather than in regional perspective. Further, with the help of limited maps and diagrams we have tried to simplify the presentation at various places.

How useful this book may prove to geographers and scholars engaged in environmental studies and what are the short comings? All these issues are of great importance. Regarding this, suggestions and criticisms will be welcomed.

It is a great pleasure to acknowledge the inestimable help we have received in writing this book from Prof. J. S. Singh, F.N.A., Department of Botany, Banaras Hindu University, Varanasi, who read the whole manuscript and made many valuable suggestions and modifications. Also we are greatly indebted to Prof. Singh for his valuable 'Foreword'.

We feel highly obliged to our teachers Prof. R. L. Singh, Prof. S. L. Kayastha, Prof. K. N. Singh, Prof. N.K.P. Sinha, Prof. R. N. Mathur and Prof. H. H. Singh for their substantial contribution by critical comments and encouragements. We are also thankful to esteemed friends Dr. H. S. Bisth, Dr. P. Nag, Dr. P. C. Pandey, Dr. S. K. Singh, Dr. M. B. Singh, Dr. H. Prasad and Shri R. S. Singh.

We record our sincere thanks to Mr. Brijesh Tiwari, District Forest Officer (Civil and Soyam), Almora, who helped us at various stages by providing useful forest statistics, without which the presentation of the book would have been tedious task. Thanks are also due to Shri A.K. Karanjia and Shri O.P. Srivastava who prepared maps and diagrams for the book, and to Shri S.D.P. Srivastava, for typing the manuscript within a record time. Shri Naurang Rai and his associates of Concept Publishing Company, New Delhi, deserve our congratulations for timely publication and efficient printing.

A. K. Singh
V. K. Kumra
J. Singh

CONTENTS

List of Figures

List of Tables

ONE

The Physical Background

Uttar Pradesh ($23^{o}52'$ to $31^{o}18'$ N and $77^{o}10'$ to $89^{o}39'$ E), the largest Indian state in terms of human population (118,862,000 in 1981) and ranking 4th in terms of geographical area (294,413 km^2), is situated in the northern part of the country. It is a land-locked state bounded by Tibet and Nepal in the north, by Himachal Pradesh in the north-west, by Haryana in the west, by Rajasthan in the south-west, by Madhya Pradesh in the south and by Bihar in the east. Till January 1950 this State was known as the United Provinces. On the map it looks like a lion in an assailing posture (Fig. 1) and is subdivided into fifty-six administrative districts.

Structurally, Uttar Pradesh, by and large, forms a major part of the Indo-Gangetic plain which lies in between the recently built Himalaya on the north and the Satpura-Vindhyan ranges of the Deccan plateau on the south. The plains is 400 km wide in its broadest part and about 2,400 km long. It is watered by the river Ganga and its tributaries and distributaries like the Yamuna, Gomti, Ghaghara, Chambal, Betwa, Ken, Rapti, Ramganga, Son (Fig. 2) and an excellent network of canals. The alluvial deposits of Uttar Pradesh (U. P.) can be categorised into: (a) old deposits known as 'Bangar' and (b) new deposits known as 'Khadar'. Generally, the Bangar land occupies higher ground and is not subjected to floods during the monsoon period. Here, carbonate of lime usually occurring in the form of nodules called 'Kankar' is preponderant. The diameter of these irregular nodules varies from 12.7 mm to 101.6 mm. The Bangar land is characterised by patches of saline and alkaline efflorescence called 'reh'. The Khadar land occupies a lower level and is generally inundated by floods during the monsoon. The 'Khadar' is most conspicuous in the eastern half of the state and forms a belt of varying width along the bank of the river Ghaghara. It neither contains 'Kankar' nor 'reh'.

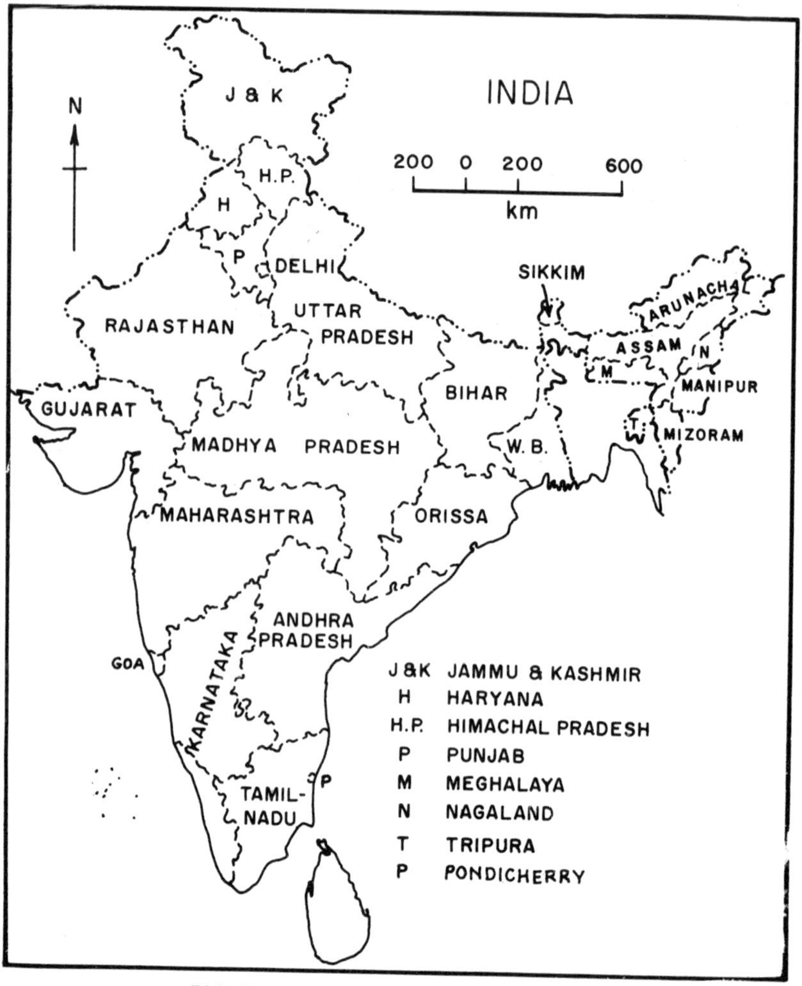

FIG. 1. LOCATION OF UTTAR PRADESH

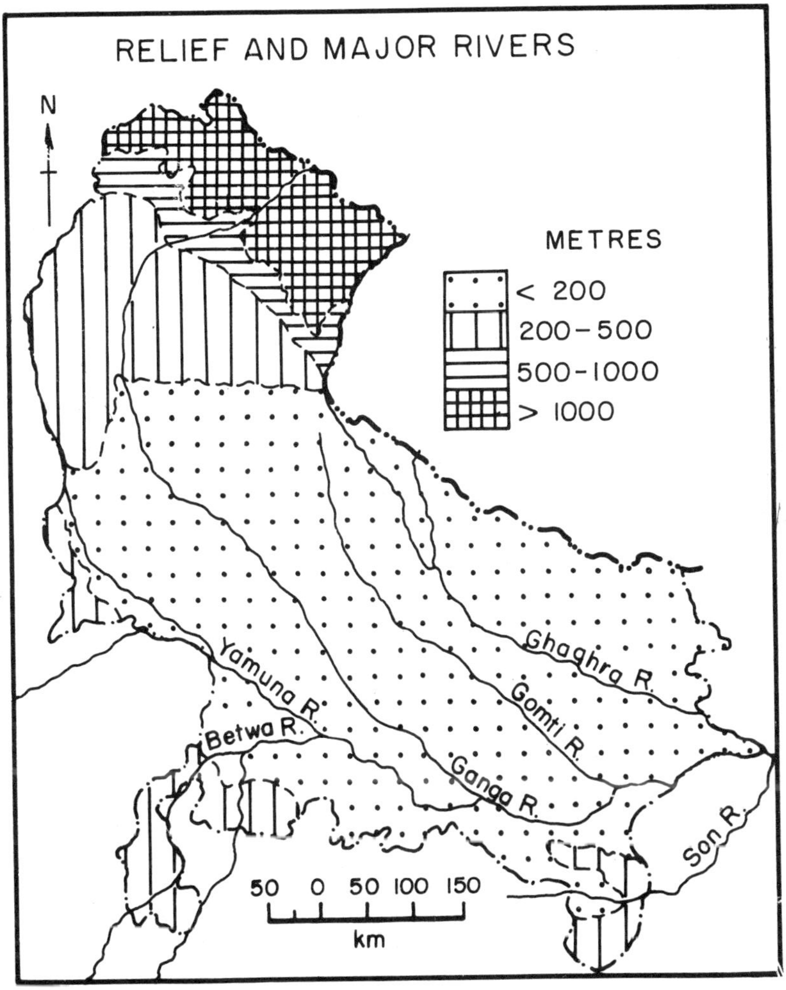

FIG. 2. RELIEF AND MAJOR RIVERS

Relief and Physiography

As Egypt is the gift of the river Nile, Uttar Pradesh is the gift of the two sacred rivers Ganga and Yamuna. On the basis of relief features, geology and drainage, the state can be broadly divided from north to south, into four distinct physiographic regions (Fig. 3): (a) the northern mountainous region (U. P. Himalaya), (b) the Sub-montane tract comprising Bhabar and Tarai, (c) the Gangetic plain and (d) the Southern plateau region.

The Northern Mountainous Region

It consists of three parallel east-west extending ranges known as Greater Himalaya, lesser Himalaya and Siwalik, in succession from north to south. The mountainous region of the U.P. Himalaya ranges in elevation from about 300 to more than 7000m (Fig. 2). It covers about 16 per cent (94,340 km^2) area of the state and comprises wholly the districts of Uttar Kashi, Chamoli, Pithoragarh, Tehri Garhwal, Garhwal, Almora and Dehradun and partly the Nainital district. The general pattern of drainage is transverse to the ranges cutting deep valleys and gorges.

The north-west portion of the state falls within the Greater Himalaya. This zone has a varying width of about 40-75 km. Its altitude ranges generally between 3000 to more than 7000 m. Except for the lower valleys, this zone is always clad with snow. Many towering mountain peaks occur, e.g., Nanda Devi (7,815 m), Kamet (7,756 m), Badrinath (7,138 m), Trishul (7,120 m), Dronagiri (7,066 m), Kedarnath (6,939 m) and Nandakot (6,861 m). Glacial landforms can be observed in fully developed forms above an elevation of 4,000 m. Some of the mighty glaciers of this region are Gangotri (25-30 km long), Milam (19 km long), Pindari, Kamet, Kedarnath and Jamnotri. From the Gangotri and Jamnotri glaciers the most sacred rivers of India - the Ganga and the Yamuna originate. These rivers have cut the landscape to make the topography highly dissected and rugged and to render the area generally inaccessible. Almost half of the total mountainous area is included in this zone.

To the south of Greater Himalaya, the lesser Himalaya extends in varying width of about 600-1000 km. Elevation varies from 1,000-3,000 m. In this part many important hill stations are situated such as Nainital (1,934 m), Almora (1,676 m), Mussoorie (2,042 m), Ranikhet (1,798 m) etc.

To the south of the lesser Himalaya the Siwalik range

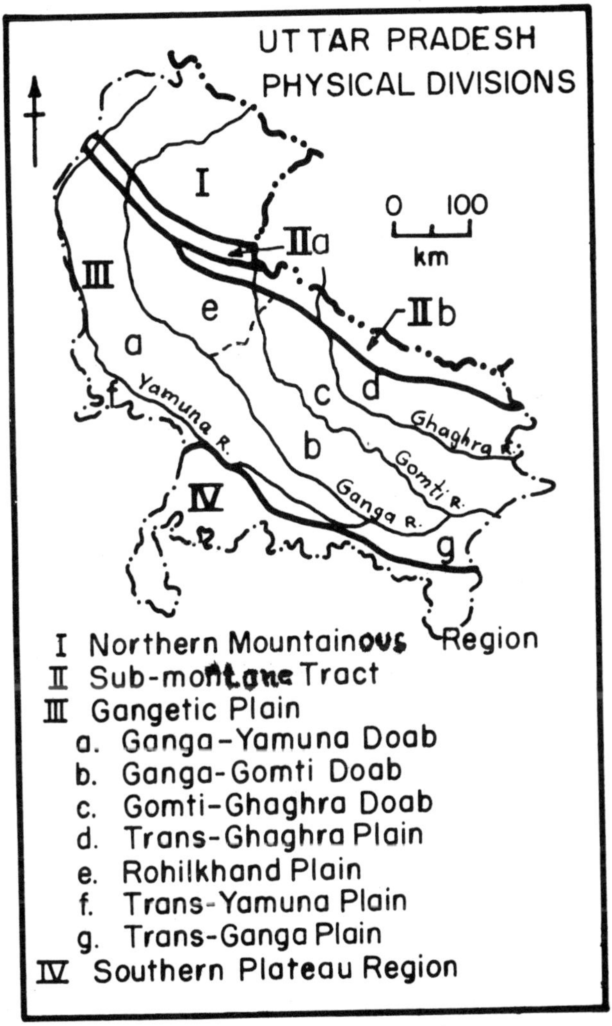

FIG. 3. UTTAR PRADESH PHYSICAL DIVISIONS

stretches. The Siwalik zone extends in a narrow belt of about 6-30 km and its height generally varies between 300 and 700 m. The Siwalik ridges are largely forested. The lesser Himalaya and Siwalik range are separated at a few places by flat longitudinal valleys called 'Duns'. Among the 'Duns' the Dun valley which is about 20-30 km wide is the most important. Other valleys of this region are Kota Dun, Patali Dun, Kothari Dun and Kriyada Dun.

The sub-montane tract

It is a transitional, narrow belt spreading in between the Siwalik and the Gangetic plain. Topographical configuration divides this region into two distinct parts: (i) Bhabar and (ii) Tarai.

The 'Bhabar' is formed by the deposition of coarse detritus boulders, gravels and pebbles and stretches in a narrow 35-40 km wide belt. The western portion of the tract is broader as compared to the eastern portion. The most striking feature of this region is that the streams and rivers flow beneath the surface. These reappear in the Tarai region. Bhabar is mostly spread over the part of Saharanpur, Garhwal, Bijnor and Nainital districts.

Extending southwards from the margins of the Bhabar, the Tarai is a wet and often swampy zone and runs parallel to the Bhabar in a narrow belt. The region receives heavy amounts of rainfall inducing swampy conditions. It is covered largely with thick forest and tall grasses, and has suffered repeatedly from the ravages of malaria and other insect-born diseases in epidemic form. The Tarai region includes the northern parts of Saharanpur, Bijnor, Bareilly, Pilibhit, Kheri, Bahraich, Gonda, Basti and Gorakhpur districts.

The Gangetic plain

It covers approximately two-third area of the state (43 out of 56 districts) and is formed by the alluvial deposits transported by the rivers Ganga and Yamuna and their tributaries - Ramganga, Gomti, Ghaghara, Sarda, Rapti and Gandak. It is a very fertile plain. At places the thickness of the deposit reaches about 4,500 m. This is not strange because a large part of this plain was a sea bed in the Post-Tertiary Period which was later filled up with Pleistocene alluvial deposits by the rivers of the Gangetic system. In general, the elevation of the plain varies from 50 to 250 m above the mean sea level. The altitude increases from east to west. The slope of the plain descends from north-west to south-east. Most of the rivers flow parallel to each other. The regions covered with old alluvial deposits (Bangar) are free from flood hazard, while the Khadars (new alluvial deposits) are liable to annual flooding and soil renewing. Many times the rivers change their course during floods and carve out entirely new channels for themselves through the Bangars. Such changes take place even in the present time. At some places ravines have been carved out due to massive soil erosion by the rivers. The ravines of Yamuna and Chambal are noteworthy. The soil of this part is the most infertile.

The Gangetic plain has been further sub-divided into five regions on the basis of the pattern of rivers running parallel to each other (Fig. 3): (i) the Ganga-Yamuna Doab, (ii) the Ganga-Ghaghara plain, (iii) the plain areas of Rohilkhand, (iv) the trans-Yamuna plain, and (v) the trans-Ganga plain.

The southern plateau region

It covers approximately as much area as covered by the mountainous region. This region is located in the extreme south of the state and is separated into two parts. The one lying in the south-west is known as Bundelkhand plateau, and the other located in the south-east is called Vindhyan plateau. In fact, it is a northerly extension of the Deccan plateau. The elevation is generally 300 m with a decline from south to north. This plateau tract is composed of older rocks such as quartzite, gneiss etc. Due to long excessive erosion the hills of this region occur as relict. The Chambal, Betwa and Cane rivers drain the Bundelkhand plateau, and the river Son and its tributaries, viz., Rihand, Bijul and Kanhar, drain the Vindhyan plateau and ultimately join the river Ganga. The Bundelkhand plateau is mostly spread over Jhansi, Lalitpur, Jalaun and Banda districts. The Vindhyan plateau extends over the most part of the Mirzapur district and Chakia Tahsil of the Varanasi district. Of the total area of the Mirzapur district - about 1,554 km^2 (13.74%) belongs to the Ganga plain (mostly the northern part of the district) and the rest falls under the Vindhyan plateau. The rivers of this region form narrow gorges and beautiful, though small falls.

Soils

Soils are very valuable natural resources which when managed skillfully continues indefinitely to supply food, fodder, fibre, timber, etc. In U. P. several types of soil are found. They can be broadly divided into the following eight categories

(Fig. 4):

Forest and hill soil

It occurs mostly in the hill districts of the state. The soil possesses comparatively coarse texture and is relatively infertile. The soils of the Himalayan region are newer and immature.

Foothill soil

It is found in a narrow piedmont zone, lying immediately below the Siwalik and extending into Saharanpur, Bijnor and Nainital districts between the rivers Yamuna in the west and the Sarda in the east. It is composed of coarser materials brought by the swift-flowing torrents from the upper mountainous coarses. The land is porous and wooded.

Tarai soil

This soil is wet and swampy. It is mostly found in parts of Nainital, Pilibhit, Bahraich, Gonda, Basti, Kheri and Gorakhpur districts.

Alluvial soil

The entire Gangetic is covered with alluvial soil deposited by the river Ganga and its tributaries. The texture of alluvial of north-west and west is lighter, of the east heavier and of the central part intermediate.

As described earlier, the alluvial soil can be grouped into Bangar and Khadar soils. Being found in comparatively higher parts, the Bangar soil is also called 'Uparhar' soil (mostly in the eastern parts of the state). Due to extensive sheet erosion at many places the soft soil has been flushed out and the Kankar and hard soil remain there. Where the percentage of Kankar is more the soil is infertile. Khadar soil is mainly confined to the flood plains. This new alluvial soil is porous, light in colour, fine textured and has a relatively higher water retaining capacity. In these soils the quantity of calcium, potash, magnesium and phosphorus fossils are more. The composition of these soils depends mainly on the silting during floods, hence each year one notices variations in soil characteristics.

Saline and alkali soil

It occurs in sizeable tract of Meerut, Ghaziabad, Buland-

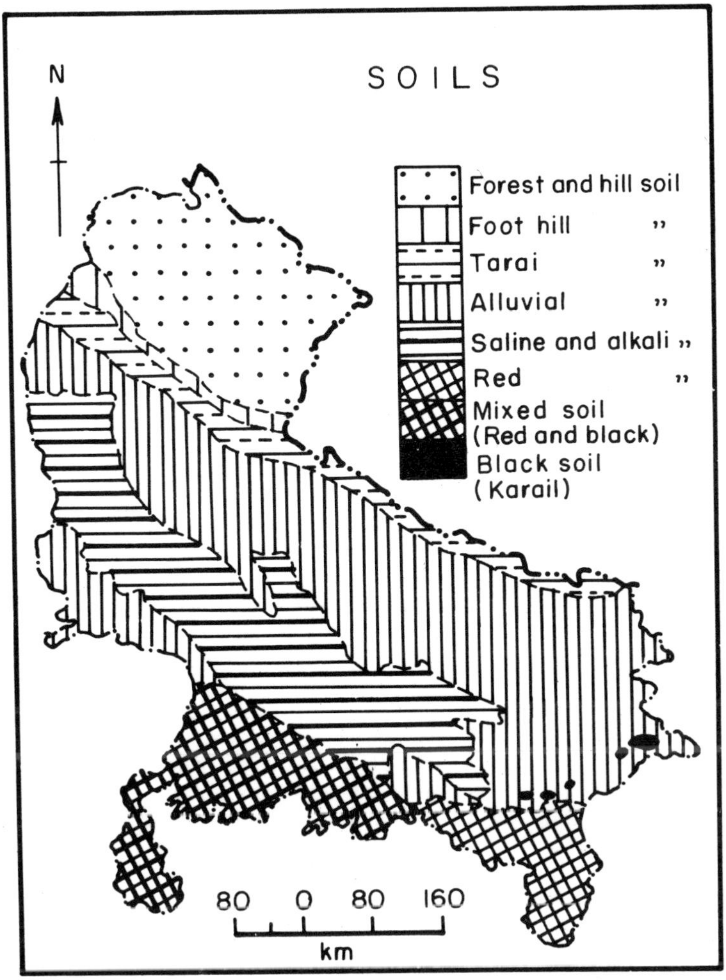
SOILS
N
Forest and hill soil
Foot hill "
Tarai "
Alluvial "
Saline and alkali "
Red "
Mixed soil
(Red and black)
Black soil
(Karail)
80 0 80 160
km

FIG. 4. SOILS

shahar, Aligarh, Etah, Mainpuri, Etawah, Kanpur, Fatehpur, Pratapgarh and Allahabad districts.

Red soil

This is confined mostly in parts of the Mirzapur district.

Mixed soil (Red and Black)

It is found in some parts of Hamirpur, Banda, Jhansi, Lalitpur and Jalaun districts.

Black soil or Karail

This soil is sticky and is found in a few localised areas of Ballia, Ghazipur and Varanasi districts.

The mountainous and plateau areas have high vegetal crown density in comparison to plains. In the remote past the invasion of population from western side was more in the Gangetic plains. Earlier it was densely wooded tract but due to anthropogenic changes the wood land was replaced by crop lands gradually. The mountainous and the plateau areas were least affected by this process probably due to harsh and inhospitable terrain, besides other factors.

Climate

Climate is one of the most important factors which influence the activities of man, natural vegetation and soil etc. In general, excepting the northern mountainous region, the climate of the state is tropical monsoonal but on account of topographical irregularities—varying altitude, latitude and distances from the sea—the climatic conditions vary from one region to another. The climatic year can be divided into three broad seasons: (i) the winter season (November to February), (ii) the summer season (March to May) and (iii) the rainy season (June to October). While the summer is pleasant in the mountainous region it is extremely hot in the plains. Similarly, the winter is characterised by bitter cold and frequent snowfalls in the mountains while in the plains it is moderate with infrequent cold waves. The plateau tracts are generally warmer in the day time on account of exposed rocky surface but the nights are comparatively pleasant.

Among the climatic factors temperature and rainfall are the most important. In the state the mean annual temperature varies from 13.5°C to 26.7°C. Excepting the foothills and hills, the temperature does not show any appreciable variation. It remains more or less the same in the plains and uplands. So far as the mean minima and maxima of temperatures are concerned the hill records lowest at Mukteshwar 17.6°C and 95°C respectively followed by Mussoorie, i.e., 10.4°C and 18°C. As we descend toward plains the temperature successively rises. In the plains Dehradun registers the lowest mean annual temperature, i.e., 21.8°C while Varanasi, Allahabad, Kanpur and Lucknow, register around 26.0°C. From the table 1.1 it is evinced that in Tarai and Western plains temperature remains around 25°C. In the uplands Jhansi records the minima and maxima as 19.7°C and 22.8°C respectively. In Tarai areas Gorakhpur registers the lowest seasonal variation in mean maximum and minimum temperatures, i.e., 9.4°C, whereas in the western plains and the southern uplands it is more than 13°C.

Table 1.1: **Showing temperature and average annual rainfall – 1931–60**

Station		Mean temperature (°C)		Mean annual temperature	Annual rainfall (mm)	Height above mean sea level
		Maxi.	Mini.			
1.	Bahraich	31.5	19.1	25.3	1177	124 m
2.	Gonda	31.5	18.8	25.1	1294	110 m
3.	Lucknow	32.3	19.4	25.8	992.4	111 m
4.	Gorakhpur	31.4	22.0	26.7	1274.4	77 m
5.	Kanpur Aerodrome	32.1	19.3	25.7	879.4	126 m
6.	Fatehpur	32.4	19.5	25.9	885.1	114 m
7.	Allahabad	32.4	19.8	26.1	1027.3	98 m
8.	Varanasi	32.2	19.8	26.0	1113.4	76 m
9.	Mussoorie	18.0	10.4	14.2	2368.1	2042 m
10.	Dehradun	27.8	15.8	21.8	2313.7	682 m
11.	Mukteshwar	17.6	9.5	13.5	1368.4	2311 m
12.	Bareilly	31.5	18.9	25.2	1068.3	173 m
13.	Aligarh	31.9	18.7	25.3	776.5	187 m
14.	Roorkee	30.2	17.0	23.6	1164.6	274 m
15.	Mainpuri	32.8	18.7	25.7	761.9	157 m
16.	Agra	32.3	19.0	25.6	765.4	169 m
17.	Jhansi	32.8	19.7	26.2	1000.1	251 m

Source: Climatological Tables of Observatories in India (1931-60), Government of India, India Meteorological Department, Poona.

Uttar Pradesh, like the country as a whole, receives most of its rainfall in summer from the south-west monsoon. The winter months bring a few showers from the north-east monsoon. The average annual rainfall in the state varies from 760 mm to 2400 mm. Table 1.1 shows the normal annual rainfall at several locations in the state. From the Figure 5 the following observations can be made: (i) the area of the lowest rainfall is encircled by a triangle formed by joining Delhi, Mainpuri and Agra in the western U.P. (below 80 cm rainfall), (ii) the Ganga-Yamuna Doab receives moderate rainfall (80-100 cm); and (iii) the areas of heavy rainfall are (a) the hilly regions with a rainfall above 140 cm, (b) the north-eastern and Tarai regions having rainfall above 120 cm, and (c) the Ganga-Ghaghra Doab and the southern border of the state having rainfall more than 100 cm.

About 85-90 per cent rainfall is received from the summer monsoon, which extends from mid-June to mid-October. As one proceeds from east to west, the amount of rainfall appreciably decreases (Allahabad: 95 cm, Agra: 62 cm) but the areas adjacent to mountainous region receive very high rainfall (Mussoorie: 240 cm, Nainital: 250 cm). In the winter season the state receives some rainfall from the north-easternly monsoon which becomes dry at the tail end of the state. The average annual rainfall of the eastern plain is 107 cm, of which, 99 cm is received from south-west monsoon. Nowhere in the state rainfall is less than 50 cm.

LAND UTILIZATION

One estimate shows that about 2/3rd of the Gross Domestic Product (GDP) comes through land resources, e.g., agriculture, forestry, minerals, etc. A knowledge of the current landuse is essential for planning rational utilization of the land resource. The pattern of landuse in an area indicates the influence and control of environmental factors, mainly of temperature, rainfall, physiography and soil.

Table 1.2 shows the land utilization pattern of the state. Of the total reporting area available for land utilization purposes, the agricultural area was 17.22 million ha (57.90%) in 1980-81. It is followed by forest (17.24%), land put to non-agricultural uses (7.66%) and fallow lands (6.33%).

The state is privileged with certain favourable climate and physical environmental factors for the development of agriculture which are denied to other neighbouring states. The fertility of the thick alluvial soil of the Ganga-Yamuna tract,

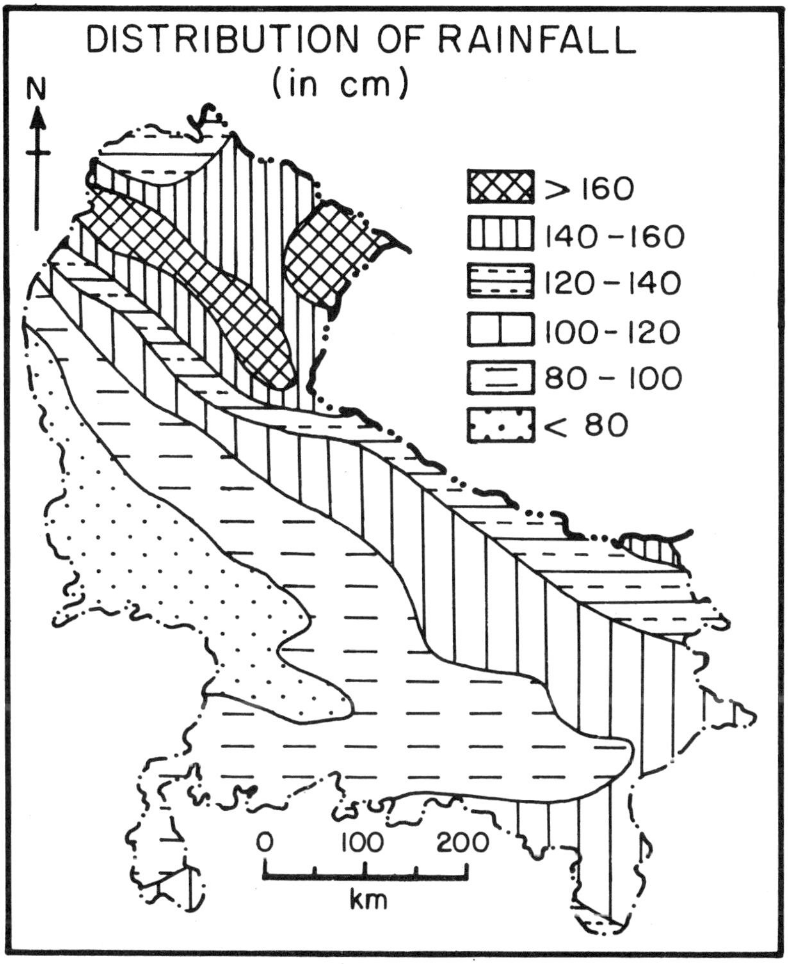

FIG. 5. DISTRIBUTION OF RAINFALL

adequate rainfall and the good network of canals and other means of irrigation have cumulatively made this state as one of the most fertile regions of the country. The main resource of the plain area is agriculture whereas the forest contributes the major share in the economy of hills and plateaus.

Table 1.2: **Land utilization pattern in Uttar Pradesh, 1980–81**

Land use categories	Area (,000 hectares)	As percentage of the total reporting area
Reporting area for land utilization purposes	29,739	100.00
Forest	5,129	17.24
Barren and unculturable land	1,141	3.83
Land put to non-agricultural uses	2,280	7.60
Culturable waste	1,148	3.86
Permanent pastures and other grazing land	296	0.99
Land under miscellaneous trees and groves	639	2.15
Current fallows	1,169	3.93
Other fallows	716	2.40
Net sown area	17,221	57.90

Source: Forest Statistics, Uttar Pradesh, 1983, Government of U.P., Lucknow.

TWO

Forest Resource

For ages the forest has shaped the life, economy and habitat of the country. Forests perform a number of functions including ecological, recreational and economic. History illustrates that for thousands of years man has recklessly destroyed forests for cultivation, habitation, industry, fuel, fodder, construction and timber, etc. The process of exploitation was relatively slow in earlier periods but with the rapid increase of population which is forming a 'j' type growth curve, it has been speeded up. Though forest in India adds a little over one per cent (1.6% in 1969-70) to the Gross National Product as compared to many developed countries of the world, there are several other indirect advantages of the forests such as flood control, ground water recharge, soil conservation, control of air pollution and of climatic extremes and surface run off and sustained precipitation, which are difficult to assess in terms of monetary benefits. On hill tops forests condense the water vapour floating in the atmosphere. Whether they always succeed in inducing precipitation or not, they invariably intercept rain to protect soil from rain beatings, soak the water and channel it to the aquifers. They pulverize rocks to make soil. Their biotic materials (leaves, etc.) contribute to topsoil formation. As being the important medium for inducing rainfall and the most vital medium for underground storage of water, the forests are our utlimate saviour from drought and water scarcity.

Thus, we can say that forests are productive, protective and aesthetic. Being the most important renewable resource, the forests, as green gold, should be managed and conserved at any cost through proper scientific measures. Because of the above mentioned facts, in any rational landuse plan for a country forests have an important place. Before proceeding to the management and conservation of forests it is appropriate to make an economic evaluation of the forest resources of a region in terms of forest area and species-wise organization of forest, wood availability, etc. In the present chapter an attempt has been made to investigate the distribution of forests and their species composition in different geographical regions. Plantations of different species raised by the Forest Department and artificial regeneration of trees have also been considered.

If we look at the distribution of forests in the country, Uttar Pradesh has a lower area under forest than many other states (Table 2.1). It has also very low per capita forest area (0.05 ha). So far as the areal coverage of forest is concerned it is spread over 17.41 per cent (51,266.55 km^2) in 1981-82 of the total geographical area of the state. However, environmentalists have expressed their doubts on the government statistics and have pointed out that even the existing forest areas are in very bad shape (Pal, 1982). Tiwari (1982) states that the official estimate of forest area includes forest blanks, scrub areas and areas above the tree limits in the Himalayan region and in consonance he felt that the wooded areas is much less than the reported area. The interpretation of landsat imageries indicates that excepting Meghalaya, Nagaland, Sikkim and Arunachal Pradesh, the remaining states and union territories have less share than the official statistics (Table 2.1). In the case of U.P. the official forest statistics depict 17.41 per cent forested area (1981-82) while the landsat imageries show only 7.14 per cent area under forest. Not only this; even the good forest accounts for no more than 6.16 per cent of the state's total geographical area. Further, the area under forest is not evenly distributed in the state. Most of the forests are confined to relatively inaccessible terrain and on high altitudes of northern mountainous and southern plateau tracts (Fig.6). The vast Gangetic plain is deprived of this valuable natural resource. The 'National Forest Policy' enunciated in 1952 laid down that the country as a whole should possess at least one-third (33.30%) of its geographical area under the forest. The mountainous and plateau tracts should have about 60.0 per cent of their area under forest, whereas the plains may have at least 20.0 per cent area under forest. Taking this norm into account we shall discuss the relative paucity or abundance of forest resources in the state.

Fig. 7 clearly reveals the uneven distribution of forest in the state. The northern mountainous region is much more heavily forested with a greater percentage of forest area than the southern plateau region. In the hill region, of the eight districts, Pauri-Garhwal, Tehri and Uttar Kashi possess more than 80 per cent area under forest whereas Almora and Dehradun have more than 70 per cent of their area under forest (Fig. 7 & 8). Among all hill districts Pithoragarh has the minimum area, i.e., 37.8 per cent under forest coverage. Singh *et al.* (1984) through the interpretation of landsat imageries have marked only 28.7 per cent forest area in the Indian Central Himalayan region (Kumaun and Garhwal region) against the official statistics of 67.3 per cent. Among it only 4.4 per cent of the area has forest with greater than 60 per cent crown density (Table 2.2). In the Vindhyan region, of the six districts, only Mirzapur possesses a significant percentage of area under forest (42.91%), whereas in the remaining districts the forest area is less than 15.0 per cent (Fig. 7 & 8). Among the Tarai region, Pilibhit and Lakhimpur (Kheri) seem to have the desired level of forest cover which is 22.44 and 21.64 per cent, respectively. It is notable that the densely populated areas of the Gangetic plain where the demand of forest resources is very high, is almost devoid of forest wealth. In the whole tract which covers thirty districts and a vast geographical area, only Varanasi has a notable percentage of forest cover (15.24%). This is because of the forests of the Chakia Tahsil which is actually a northerly extension of the Vindhyan uplands. As many as eleven districts have less than one per cent and eight districts have 1-2 per cent area under forest (Table 2.3). This indicates the grave situation with regard to this nature-gifted wealth.

It is notable that of the fifty-six districts in the state, as many as twenty have only a nominal share of their area under forest (12%). Among these, the four districts of the eastern U.P., viz., Ballia, Ghazipur, Jaunpur and Azamgarh, have hardly any forest tract worth the name.

So far the suggested national norm is concerned, the state as a whole does not possess the requisite percentage of forest area. If we take the national norm for different terrain it is interesting to note that except for Pithoragarh, all the hill districts more or less have the requisite percentage, whereas none of the districts of the plateau region fulfils the national norm. The situation worsens in the Tarai and Gangetic plain. Here excluding Pilibhit and Lakhimpur, all districts posses very little forest area compared with the 20 per cent level adopted in the National Forest Policy.

Originally there was a large area under the forest but with the rapid increase in population more area was cleared for cultivation and habitation. On account of rough terrain and high altitudes it was difficult to tap the forest resources of the mountainous region and that of the extreme south easterly district of the state, viz., Mirzapur, hence the axe of man fell down mercilessly on the forests of the vast Gangetic plain which is relatively a fertile and homogeneous area. Because of this, most of the area of the Gangetic plain was utilized for agriculture and settlements.

Table 2.1: **State-wise area under forests**

State/Union Territory*	Forest area as percentage of total geographical area in 1980	Forest area as percentage of total geographical area	Per capita forest area (in ha)
Andhra Pradesh	23.15	14.61	0.13
Assam	39.11	25.21	0.16
Bihar	16.81	11.58	0.04
Gujarat	9.96	2.58	0.10
Haryana	3.71	0.91	0.01
Himachal Pradesh	38.06	16.40	0.51
Jammu & Kashmir	9.85	6.46	0.37
Karnataka	19.75	13.88	0.11
Kerala	28.62	18.98	0.04
Madhya Pradesh	34.75	20.37	0.29
Maharashtra	20.82	9.86	0.11
Manipur	67.75	60.70	1.06
Meghalaya	38.02	55.39	0.66
Nagaland	17.42	48.97	0.41
Orissa	43.44	25.31	0.25
Punjab	4.83	0.99	0.02
Rajasthan	10.20	1.75	0.11
Sikkim	35.62	39.49	1.04
Tamil Nadu	16.75	10.14	0.05
Tripura	56.58	49.46	0.29
Uttar Pradesh	17.46	7.14	0.05
West Bengal	13.48	7.78	0.02
Andaman & Nicobar Islands	86.13	77.49	3.90
Arunachal Pradesh	61.67	62.34	8.26
Dadra and Nagar Haveli	40.82	25.51	0.23
Goa, Daman & Diu	27.56	29.90	0.09
Mizoram	33.81	56.76	1.46
India	22.73	14.10	0.11

Source: (1) India's Environment, 1982.
(2) National Remote Sensing Agency (NRSA), Hyderabad, 1983.

* Chandigarh, Delhi, Pondicherry and Lakshadweep have negligible share.

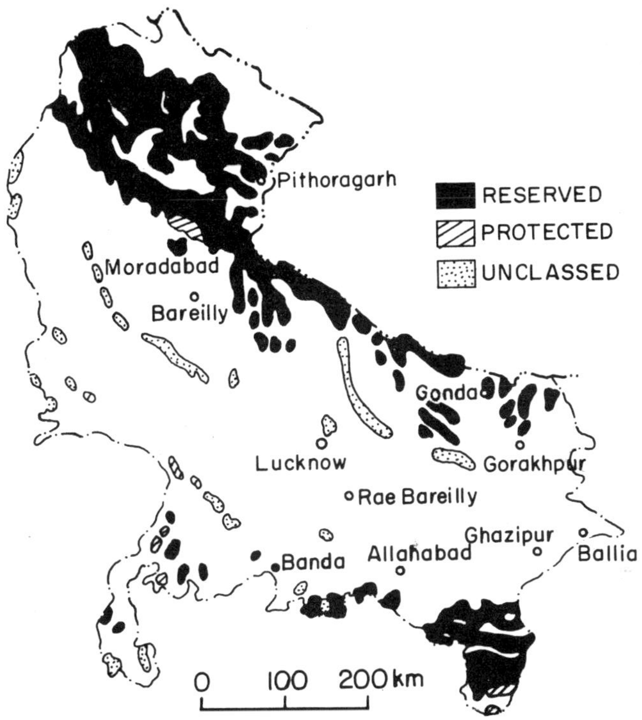

FIG. 6. SPATIAL DISTRIBUTION OF FOREST

Considering the population on the one hand and the area under forest on the other, the per capita forest area for all the districts lying in different geographical regions, has been calculated to illustrate the pressure of population on forest land. The per captia forest area amounts to 0.046 ha in 1981-82 compared with 0.110 per person which is the average for whole India. Among the districts of U.P., Uttar Kashi, registers the highest per capita forest land, i.e. 3.72 ha, followed by Chamoli (1.45 ha), Tehri (0.80 ha), Pauri (0.71), Pithoragarh (0.67 ha) and Almora (0.52 ha). Thus, the per capita forest area tells the same story as the geographical area under forest. It is obvious that densely populated areas with meagre amount of forest cover have low per capita forest land, whereas the thinly populated area of the hilly and plateau tracts have a higher per capita forest area.

Composition of Forests

From the economic point of view the accessibility and the composition of forest (type of wood) are more important than

Table 2.2: **Lanuse and forest cover in the Indian Central Himalaya interpreted from satellite imageries**

Landuse	Kumaun		Garhwal		Total	
	Area (km^2)	Per cent of total area	Area (km^2)	Per cent of total area	Area (km^2)	Per cent of total area
Non-forest						
Snow	2,968	14.2	4,973	16.6	7,941	15.6
High altitude meadows	732	3.5	1,108	3.7	1,840	3.6
Other non-forest	10,139	48.2	12,670	42.3	22,789	44.7
Total non-forest	13,819	65.9	18,751 +3,744*	75.1	32,570 +3,744*	71.3
Forest						
Poor forest	1,767	8.4	2,942	9.8	4,709	9.2
Medium forest	4,381	20.9	3,285	11.0	7,666	15.0
Good forest	1,017	4.8	1,246	4.1	2,263	4.4
Total forest	7,165	34.1	7,473	24.9	14,638	28.7

* Imageries were not available for 3,744 km^2 area in Garhwal but this area is coverdmostly by snow and altitude meadows.
Source: Singh, J.S. et al. (1984), AMBIO, Volume 13(2).

the mere knowledge of area under forest because each species requires a particular geographical environment, and has varied economic impact. For example, in the Government's reforestation programme, particularly in ecologically threatened areas, the fast growing exotic species - Philippinian/Australian Eucalyptus has been vigogously planted, but the species is shrouded in controversies ever since it became a part of the government's afforestation programme. In a panel discussion on 'choice of tree species for different parts of India' organized by the Society for Promotion of Wasteland Development (SPWD) in December 1984, Dr. T.N. Khoshoo, the then Secretary, Department of Environment, was firm in his view that Eucalyptus should be barred from conservation of forestry. He had no objection to this exotic tree being planted for production of forestry. But Dr.K.M. Tiwari, President, F.R.I., Dehradun claimed that there was no scientific evidence to prove that Eucalyptus hastened desertification of land or was responsible for lowering the ground water table thus degrading the land. But in any case it is beyond doubt that fast growing species will naturally consume more ground water and nutrients. There are certain species

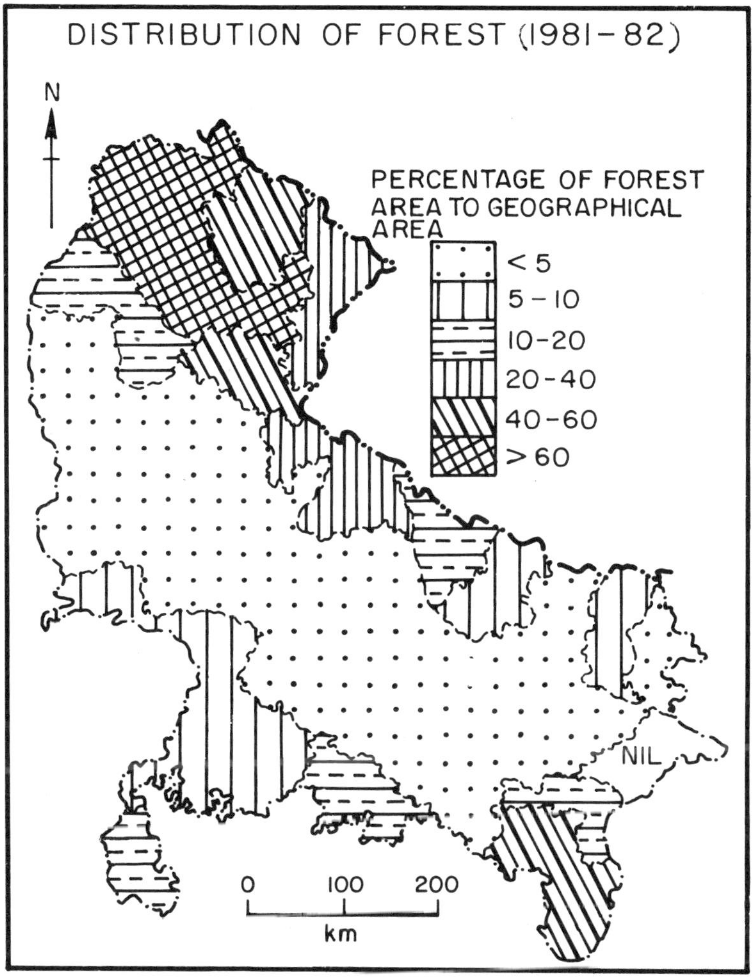

FIG. 7. DISTRIBUTION OF FOREST (1981-82)

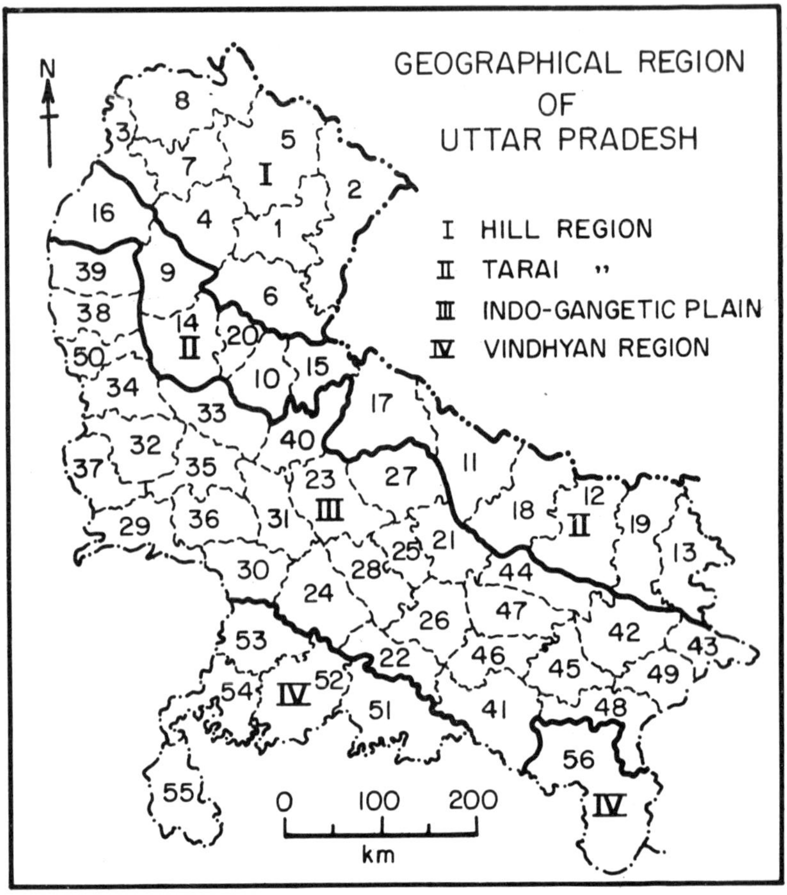

FIG. 8. GEOGRAPHICAL REGION OF UTTAR PRADESH

1. Almora	2. Lithorgarh	3. Dehradun	4. Garhwal (Pauri)
5. Chamoli	6. Nainital	7. Tehri (Garhwal)	8. Uttar Kashi
9. Bijnaur	10. Bareilly	11. Behraich	12. Basti
13. Deoria	14. Muradabad	15. Pilibhit	16. Saharanpur
17. Lakhimpur(Kheri)	18. Gonda	19. Gorakhpur	20. Rampur
21. Barabanki	22. Fatehpur	23. Hardoi	24. Kanpur
25. Lucknow	26. Raiberely	27. Sitapur	28. Unnao
29. Agra	30. Etawah	31. Farrukhabad	32. Aligarh
33. Badaun	34. Bulandshaher	35. Etah	36. Mainpuri
37. Mathura	38. Meerut	39. Muzaffarnagar	40. Shahjahanpur
41. Allahabad	42. Azamgarh	43. Batha	44. Faizabad
45. Jaunpur	46. Pratapgarh	47. Sultanpur	48. Varanasi
49. Ghazipur	50. Ghaziabad	51. Banda	52. Mamirpur
53. Jaloun	54. Jhansi	55. Lalitpur	56. Mirzapur

Table 2.3: **Distribution of forest area by District/Geographical Region (1981–82)**

	District/ Region	Geographical area (in km²)	Area under forest (in km²)	Percentage of forest area with respect to geographical area	Population in 1981 (in 000)	Per capita forest area (in ha)
1.	Almora	5385	3944.32	73.25	757	0.521
2.	Pithoragarh	5856	3302.89	37.30	489	0.675
3.	Dehradun	3088	2225.14	72.06	762	0.292
4.	Garhwal(Pauri)	5440	4656.12	83.75	638	0.714
5.	Chamoli	9125	5269.36	57.74	364	1.448
6.	Nainital	6794	4026.40	59.26	1137	0.345
7.	Tehri-Garhwal	4421	3972.00	89.84	498	0.798
8.	Uttar Kashi	8016	7102.77	88.61	191	3.719
	Hill Region	51125	34399.00		4836	
9.	Bijnaur	4849	708.02	14.60	1839	0.039
10.	Bareilly	4120	4.11	0.10	2273	0.0002
11.	Behraich	6877	1017.73	14.80	2216	0.046
12.	Basti	7229	82.34	1.14	3578	0.002
13.	Deoria	5445	11.51	0.21	3497	0.0003
14.	Moradabad	5967	119.21	2.00	3149	0.004
15.	Pilibhit	3499	785.03	22.44	1008	0.078
16.	Saharanpur	5591	773.15	13.83	2674	0.029
17.	Lakhimpur(Kheri)	7680	1661.60	21.64	1953	0.085
18.	Gonda	7352	718.24	9.77	2835	0.025
19.	Gorakhpur	6273	556.30	8.87	3796	0.015
20.	Rampur	2367	66.11	2.79	1179	0.006
	Tarai Region	67249	6503.35		29997	
21.	Barabanki	4401	78.34	1.78	1992	0.004
22.	Fatehpur	4152	48.64	1.17	1572	0.003
23.	Hardoi	5986	111.73	1.87	2275	0.005
24.	Kanpur	6176	118.62	1.92	3742	0.003
25.	Lucknow	2528	94.72	3.75	2014	0.005
26.	Raibarely	4609	48.02	1.04	1887	0.003
27.	Sitapur	5743	43.28	0.75	2337	0.002
28.	Unnao	4558	158.29	3.74	1823	0.009
29.	Agra	4805	394.42	8.21	2853	0.014
30.	Etawah	4326	386.83	8.94	1743	0.022
31.	Farrukhabad	4274	32.70	0.77	1949	0.002
32.	Aligarh	5019	8.08	0.16	2375	0.003
33.	Badaun	5168	69.03	1.34	1972	0.004
34.	Bulandshahar	4352	103.70	2.38	2358	0.004
35.	Etah	4446	11.66	0.26	1859	0.001
36.	Mainpuri	4343	65.74	1.51	1726	0.004
37.	Mathura	3811	15.75	0.41	1560	0.001
38.	Meerut	3911	79.93	2.04	2767	0.003
39.	Muzaffarnagar	4176	72.76	1.74	2275	0.001
40.	Shahjahanpur	4575	111.77	2.44	1848	0.006
41.	Allahabad	7261	201.42	2.77	3797	0.005
42.	Azamgarh	5740	2.38	0.04	3544	0.0001
43.	Ballia	3189	-	-	1954	-
44.	Faizabad	4512	12.68	0.28	2382	0.001
45.	Jaunpur	4038	2.16	0.05	2533	0.0001
46.	Pratapgarh	3717	4.44	0.12	1801	0.0001
47.	Sultanpur	4436	24.97	0.56	2443	0.001
48.	Varanasi	5092	774.04	15.24	3701	0.021
49.	Ghazipur	3377	-	-	1945	-
50.	Ghaziàbad	2590	25.25	0.99	1843	0.001
	Gangetic Plain	135,309	3107.87		68,861	
51.	Banda	7624	770.81	10.24	1534	0.051
52.	Hamirpur	7166	273.18	5.21	1194	0.031
53.	Jalaun	4565	257.31	5.64	986	0.026
54.	Jhansi	5024	325.44	6.48	1137	0.029
55.	Lalitpur	5039	669.95	13.30	578	0.116
56.	Mirzapur	11310	4852.62	42.91	2039	0.238
	Vindhyan Region	40728	7256.31		7468	
	Uttar Pradesh	2,94,413	51,266.53	17.41	1,10,862	0.046

which reduce the levels of pollution and check the sedimentation. Even though we have 3,000 species of forest trees and 300 species of timber, no concerted research has been taken up to select native fast growing species for reforestation. Regardless of their suitability to our environment the country is heavily dependent upon exotic fast growing species.

Species like Shahjana, Khejari, Bar, Babool (Acacia) and various other acacia varieties are suitable for various types of lands in India. Khejari (Prosopies) checks the desertification. Dr. V. Shankar has pointed out that at places where river water had been rendered turbid or the flow was too fast, causing tremendous soil erosion and silting, massive and discriminate plantations could be of help. The most effective and natural process would be to plant suitable species at suitable sites. For instance water turbidity can be reduced by planting certain water plants such as 'water-hyacinth' which has the capacity to absorb 25-30 types of minerals. Again the flow of industrial effluents could be slowed down at strategic points by 'water-hyacinth' which is a very fast growing and wide spreading plant. However, water hyacinth is a notorious weed, difficult to eradicate, and its use for such purposes must be thoroughly studied before any general recommendation is made. Trees like Katha, Shisham, Shemal, Jamun etc., should be planted in flood affected areas because they can tolerate submergence.

Replantation has checked the heavy sedimentation and pollution caused by rock phosphate mining done in Mahadeota region since 1960 in the perennial Bandal river which provides drinking water to Dehradun city. It is advisable to plant different species in different layers to arrest over-burden and to stabilize rain water courses along them. In addition, it has been proved that there is a considerable difference in air temerature under canopies of different species and open places. (Dabral et al., 1969). It has been observed that the oak forests have played a great role in maintaining the spring environment in Kumaun and Garhwal Himalaya. Since a significant portion of human settlement is located in oak zone of Kumaun and Garhwal Himalaya, the role of oak forest in maintaining the general environment and life of spring is significant (Singh, 1981). The depletion of oak forest in the Kumaun and Garhwal Himalayan region has thwarted the very existence of natural springs. The areas covered with oak forests are more moist than those by chir-pine and certainly oak forests tend to increase the amount of soil water by reducing surface run-off, by increasing the permeability of soil, by decreasing evaporation and by increasing water holding and storage capacity of the soil (Divedi, 1980).

Thus, replacement of oak forest by species like chir-pine is not desirable otherwise it will create a new problem of rural bio-geocenosis.

Composition of forest refers to different plant species found therein. In Uttar Pradesh the forests can be categorised into two groups, i.e., coniferous (soft wood) and non-coniferous (broad leaved having hard or soft wood). Coniferous forest is one in which 80 per cent or more of the volume is contributed by coniferous species and non-coniferous is that in which 80 per cent or more of the volume is of non-coniferous species. The forests of U.P. Himalaya mainly consits of sal in the foot hills, pine in the middle altitudinal range and oak at altitudes of 1900 m and above. Because of bio-edaphic factors and certain nutrient conserving strategies pine has tended to spread vigorously at the expense of oak (Singh et al. 1984). Gupta (1979) has reported that in the hilly region, forests are mainly coniferous-chir pine being the main plant species along with some deodar and fir forests. The oak and broad leaved species are interspersed with coniferous forests and occupy more favourable areas with better soil and moisture conditions.

In U.P., out of 50,750 km^2 of forest area, coniferous species cover 12,530 km^2 and the remaining area is stocked with broad-leaved forests. Among the non-coniferous or broad leaved forests sal, ashna, teak, khair, sisham, semal, haldu, jamun, eucalyptus, oak etc., are the important species. So far the area under different species is concerned, chir is found over 10.12 per cent (411,782 ha) of the total forest area under the control of forest department followed by fir and spruce: 2.02 per cent, Deodar and Kail 0.45 per cent each. The cypress is found just only on 3,295 ha, i.e., 0.08 per cent of the total forest area (Table 2.4). The wood of coniferous forests are soft and light in weight which makes them easier to cut and transport. The broad leaved species are mostly hard wooded and several of them are of high economic value. Among the broad leaved species, sal occupes 13.16 per cent (535,287 ha) of the total area followed by teak (1.27%), khair and Shissoo (1.83%). When these species are found intermingled with other species, then the extraction of timber becomes relatively a costly affair on account of the problem of the selectivity and further to bring them in the market. These mixed species altogether occupy the largest area of the forst, i.e., 46.89 per cent. In addition, 23.55 per cent of the forest area is covered by species which are unproductive from the economic point of view. So far the outturn of timber is concerned, the non-coniferous (574,034 cu.m. in 1981-82) as compared to the coniferous species (373,078 cu.m.).

Table 2.4: **Composition of forest areas under different species**

Species	Area in (as on 31.3.82) hectare	As percentage of the total forest area
Chir	411,782	10.12
Deodar	18,411	0.45
Fir and spruce	89,405	2.20
Kail	18,417	0.45
Cypress	3,295	0.08
Sal	3,35,287	13.16
Teak	51,792	1.27
Khiri Sissoo	74,429	1.83
Miscellaneous species	1,907,780	46.89
Unproductive and blank	958,120	23.55
Total	4,068,718	100.00

Source: Forest Statistics, Uttar Pradesh, 1983.

Distribution of Forests by Legal Status, etc.

For the convenience of administration, forests have been grouped into three broad categories: (i) State forest which is under the direct control of the government and revenue earned goes to the state government, (ii) the community forest which is under the control of autonomous bodies and the Panchayat etc., and (iii) Private forest - it is under private ownership and usually includes plants which provide fruits and timber for construction as well as fuel wood. Further, on the basis of the legal status, the forest department has categorised forests into three broad divisons: (i) reserved, (ii) protected and (iii) unclassed and vested. Reserved forest is one which is permanently dedicated either to production of timber or other forest produce and in which the right of grazing is seldom allowed. It is owned and operated by the government. In protected forest these rights are allowed subject to a few mild restrictions. Though owned by the government, it may be used by private individuals to meet their needs of wood and grazing. The unclassed and vested forests are those which are under the general control of Forest Department and are allowed to be used by general public.

In addition to these, there are other forest categories also such as civil forest (under the control of revenue authorities), soyam forest, panchayat forest, and municipal and contonment forest etc. The areal distribution of forest by legal status is given in table 2.5. From the table 2.6 it is evident that about 79.36 per cent (40,687 km^2) forest area is under the supervision and control of the Forest Department. Among this, 85.58 per cent is the reserved forest while protected and unclassed forests cover 4.39 and 10.02 per cent area, respectively. The remaining 20.64 per cent of the total forest area is covered by civil forests mainly confined in Almora, Pithoragarh, Garhwal, Chamoli and Nainital districts; by soyam forests: Uttar Kashi, Tehri and Dehradun districts; by Panchayat forests: Almora, Garhwal, Chamoli, Nainital and Pithoragarh districts; by Private forests: Dehradun, Almora; by Municipal and cantonment forests: Dehradun Nainital and Garhwal districts.

Of the total forest area under the control of the forest department, the hills possess 58.54 per cent, while the rest occurs in the plains. In the hilly tract almost cent per cent area (99.6%) has been declared as reserved forest because the area is ecologically more sensitive. Any damage to forests of the hills means a real disaster to the plains. Only 0.26 and 0.13 per cent area is under protected and unclassed forests, respectively. The plains have 65.78 per cent of the forest area under the category of reserved forests while 10.23 per cent is protected and 23.98 per cent is unclassed forests. In hills on account of unirrigated conditions, small holdings, and severity of climate, agriculture is not sufficient to provide fodder to the livestock. Hence, the natural vegetation acts as a fodder resource for the animals. Over-grazing ultimately breaks the surface which leads to massive soil erosion and consequently the siltation of rivers and dams. The other associated problems consequent to misuse of forest lands are landslides and avalanches, etc. Because of these problems, the goverment has classed the forests of Hill districts under the cateogry of reserved forest. It will lessen the menace posed on the natural environment.

Forest Types

The forest classification followed in this book is mainly based on the work of Champion and Seth (1968). Climatic factors particularly temperature and rainfall, soil and physiognomy are the basis for dividing the forests into various types. Thus, the forests of the State can be set out into the following categories (Fig.9):

(a) The tropical deciduous forests:
  (i) Moist deciduous and
  (ii) Dry deciduous,
(b) The tropical thorn forest,
(c) The sub-tropical pine forest,

(d) Temperate forest and
(e) The alpine scrub and forest.

Table 2.5: **Distribution of forest area in hills and plains by legal status (1981-82)**

Category	Area (in km^2)	As percentage to total area under forest
1. Reserved	34,822.75	85.58
(a) Hills	23,726.57	58.31
(b) Plains	11,096.18	27.27
2. Protected	1,787.59	4.39
(a) Hills	61.35	0.15
(b) Plains	1,726.24	4.24
3. Unclassed and vested	4,076.84	10.02
(a) Hills	31.73	0.08
(b) Plains	4,045.11	9.94
Total	40,687.18	100.00

Table 2.6: **Distribution of forest in U.P. under the control of various agencies (1981-82)**

Category	Area (in km^2)	As percentage of total forest area
Forest area under the control of forest department	40,687.18	79.36
Other forests	10,579.35	20.64
(i) Civil forests under the control of revenue authority	6,072.00	11.84
(ii) Soyam forests	1,941.63	3.78
(iii) Panchayat forests	2,368.00	4.63
(iv) Private forests	158.88	0.31
(v) Municipal and cantonment forests	24.31	0.05
(vi) Other forests	14.53	0.03
Total	51,266.53	100.00

The Tropical Deciduous Forest

This type of forest is found in those regions where the annual rainfall varies from 50-200 cm. On the basis of the amount of rainfall this can be further divided into two sub-types : (i) Moist deciduous forest. In the state these forests are found below the Siwalik ranges particularly in the Tarai and Bhabar regions. These two sub-montane tracts receive average rainfall between 100-200 cm. This type of forest is stretched over an area of 6,21,975 ha, i.e., 15.28 per cent (Table 2.7). Sal and teak are the most important species of this forest type besides haldu, gutel, khair, kanju, semal, sissoo, tun etc. (ii) Dry deciduous forest : This type of forest is found in those areas where the annual rainfall varies between 50-100 cm. On account of low rainfall, the height of the trees is lower. They occupy nearly one-third forest area (33.23%) of the state. Shisham, haldu, khair, siris, tendu, chandan and mahua are the important trees of this forest type. Sal and teak are also found in restricted areas. This type of forest is mostly found in a compact zone in the lower valleys besides the southern plateau region. Though the wood of this forest is of poor quality, its commercial utilizarion is more.

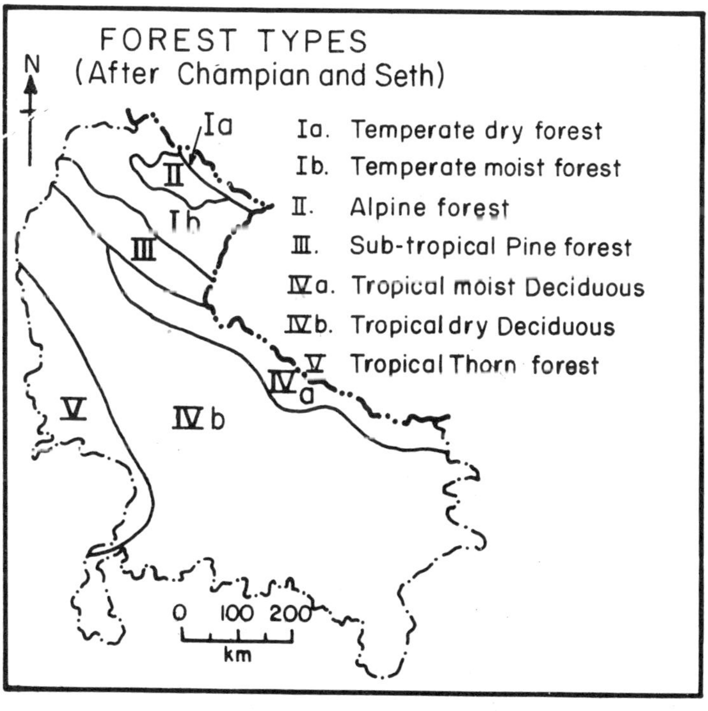

FIG. 9. FOREST TYPES

Table 2.7: **Area by forest types (1981-82)**

Forest types	Area (ha)	As percentage of forest area under the control of forest department
Tropical moist deciduous forest	6,21,975	15.28
Tropical dry deciduous forest	13,52,132	33.23
Tropical thorn forest	71,811	1.76
SubTropical pine forest	4,21,979	10.37
Himalayan dry temperate forest	2,979	0.07
Himalayan moist temperate forest	5,79,003	14.23
Sub-alpine forest	37,437	0.92
Moist alpine scrub forest	1,19,748	2.94
Dry alpine scrub	62,678	1.54
Forest area under the control of forest department	40,68,718	

The tropical thorn forest

It is concentrated in the dry south-west region of the state where the rainfall is less than 50 cm and includes babool and other similar varieties of plants. Due to very low amount of precipitation the trees are small and often thorny with small leaves and deep roots. This category of forest covers an area of about 1.67 per cent. They occupy sites that are otherwise useless from the viewpoint of animal husbandry and fuel. The timber of these forests is useful. The bark of Babool is utlilized in vegetable tanning.

The sub-tropical Pine forest

It occupies an area of 4,21,979 ha (10.27%). It is found between 900-1800 m altitude and is confined to Nainital, Almora and Pithoragarh districts. Chir is the most valuable and important species of this forest. The average height of chir varies in between 20-30 m. The timber is of multipurpose and is extensively utilized for tea packing. In addition it yields resin which is a very important commercial commodity.

The temperature forest

The temperate forest has been further divided into two subclasses: (i) The moist temperate forest: this type of forest is mostly found in the Himalayan region of Nainital, Almora and Pithoragarh districts at 1800-2500m altitude. This is also a region for various type of conifers such as deodar, fir and spruce, blue pine and cypress. Deodar is mainly found in those regions where the annual rainfall varies in between 100-200 cm. The timber of deodar is generally hard and durable. The timber of fir and spruce is mostly used for packing cases and as a raw material for paper and pulp and match industries. In spite of various economic uses it is very difficult to tap these resources because of their existence in remote and relatively inaccessible mountainous zones. This type of forest covers an area of 579,003 ha. (14.22%). (ii) The dry temperate forests: These are found over an area of 2,979 ha in the inner ranges of the Garhwal Himalaya. Deodar is the most important species of this forest.

The Alpine Scrub and Forest

These are found in the highest altitudinal zone of the Himalaya (2800-3600 m) and occupy an area of 2,19,971 ha. In alpine forest the trees of temperate forest are mainly found. The timber of these trees is soft but durable. But because of limited accessibility it is difficult to exploit this forest for timber. On the basis of altitude this has been divided into two sub-groups: (i) sub-alpine forest: It is found in altitudinal zone of 2800-3050 m. This type of forest is spread in the upper valleys of narrow mountainous ranges as well as in the central region of Nainital. Oak, blue pine, cypress and deodar are the important species of this type of forests. (ii) Alpine scrub forest: It is found at an elevation of a little over 3,050 m in Nainital, Pithoragarh and Almora districts and occupies an area of 182,534 ha (4.58%). The vegetation include scrubs like Rhododendrun, Campanulatum, pryus-foliolsa and Juniperous-recurva. The moist alpine scrub forest is found over an area of 1,19,739 ha whereas the dry alpine scrub forest is found in an area of 62,786 ha.

In addition to the above forest types there is an area of 7,64,498 ha (18.79%), of which 6,00,710 ha are blank snow covered, rocky and wasteland and the remaining are under plantation and unclassified forests.

Plantation

Despite favourable geographical conditions, the area under forest is only 7.14 per cent (NRSA, 1984) of the total area of the state as against the national norm of 33 per cent. It reflects the poor state of affairs in matters of natural vegetation. Though forests are the inexhaustible and renewable resources, the indiscriminate felling of trees specially after the advent of Britishers has played havoc with this national heritage. There are many causes which have contributed to the destruction of forests. Among the causative factors, various activities related to "development" such as multipurpose river valley projects, construction of roads and railways, extension of area under cultivation and habitation, establishment of industries and townships and construction related to various schemes are most significant. The other notable detrimental factors are the forest fires, overgrazing, excessive extraction of timber by contractors and unsocial elements, etc. This has resulted in environmental degradation in terms of flood, soil erosion, siltation, landslides, drought, lowering of the ground water table, increasing rate of evaporation, change in micro climate, etc. Therefore, to maintain the ecological balance and to meet the needs of ever increasing population, it is desirable and essential to have the requisite area under forest taking into account the nature of the terrain. In the area under study most of the forest is concentrated in the Kumaun and Garhwal divisions whereas the people of the plains are deprived of this resource. The situation is so grave that they are unable to even fulfil their day to day domestic requirements. There is a vast area lying in the plains such as ravines, barren and wasteland. This as well as the area which has already been denuded in mountainous regions should be utilized for afforestation purposes on a war footing. It has been attempted to investigate the progress made in this regard in different successive plan periods.

Progress made in regeneration and afforestation

In 1978-79 an area of 5,045 km^2 was brought under the programme of regeneration and afforestation. Out of this, regeneration treatments were carried out on about 4968 km^2 by the Forest Department itself. In the regeneration and afforestation programmes, the maximum area (4606.07 km^2) was selected for natural regeneration while 134.12 km^2 area was regenerated by Coppice*. About 228 km^2 area was utilised for artificial regeneration (Table 2.8). Of this, 60.38 km^2 area was under the existing forest and 167.16 km^2 area was devoted to reforestation. In addition, on 77.37 km^2 area which is not under the control of forest department, afforestation programme was also successfully carried out. It reflects the people's cooperation and awareness towards this programme.

In 1979-80, 1980-81 and 1981-82, respectively about 4,605;

---

* Coppice: wood of small trees grown for periodical cutting.

4,632 and 4,760 km^2 area was brought under regeneration programmes. In these periods the cost of regeneration operation has increased tremendously from Rs.92.70 million (1979-80) to Rs.203.14 million (1981-82) because of inflation.

Table 2.8: **Process made in concentrated regeneration and afforestation (in ha)**

Item	1978-79	1979-80	1980-81	1981-82
1. Area under concentrated natural regeneration	4,60,607	4,13,050	4,10,308	4,15,277
2. Area regenerated by coppice	13,412	9,222	13,378	10,293
3. Area artificial regenerated:				
(a) In existing forest	6,038	5,458	6,062	8,563
(b) Afforestation	16,716	15,667	12,112	15,569
Total area artificially regenerated under the control of forest department	22,754	21,125	18.174	24,132
Afforestation carried out in the area not under the control of Forest Department	7,737	17,183	21,379	26,304
Total cost of regeneration (in million of rupees)	72.35	92.70	120.00	200.34

Remarkable progress in artificial regeneration has been made from 1930-31 to 1980-81. In the year 1930-31 artificial regeneration was carried out just over an area of 3.57 km^2. The artificial regeneration was made over an area of 20.62 km^2 in 1950-51 whereas in 1960-61 it went up to 154.53 km^2. It has further increased to 304 km^2 in 1970-71. Seeing the massive environmental damages caused by deforestation, the government paid more attention towards afforestation, and in accordance allocated more funds to achieve the desired goal. Consequently, in 1981-82 the area under artificial regeneration increased to 504.36 km^2 (241.32 km^2 under the control of Forest Department and 263.04 km^2 outside of this control).

It has been observed that so far in the regeneration programme about Rs.735.50 million (1930-31 to 1981-82) have been spent. Therefore, care should be taken not to strip off this regenerated area again. If it happens then this will be a callous waste of public funds. This aim can be achieved only if the people treat forests as long-term valuable crop and then only we will be saved from the massive environmental degradation caused by the reckless cutting of trees in the name of development.

Species-wise Plantation

Plantation does not mean only to plant trees for number and extent but it is rather more important to plant trees according to their economic, ecological and biological suitability. For instance, in area, of oak forest which maintains the spring-environment, other species like chir should not be planted. Likewise a fast growing species, viz., Eucalyptus should not be planted in ecologically sensitive areas. Besides, considering the favourable environmental conditions plantation of suitable species should be raised at appropriate sites. Looking into the above referred facts it has been attempted to analyse the progress of plantation made species-wise. In 1980-81, the forest department artificially regenerated different species on 241.32 km^2 land. Of this, coniferous species were developed over an area of about 15.81 km^2 and non-coniferous species on 225.51 km^2. Among the non-coniferous species, teak occupies the maximum area (39.89 km^2) closely followed by eucalyptus (31.56 km^2). The statistics reveal that right from 1978-79 to 1981-82 the plantation area has increased under species like teak, eucalyptus, semal, gutel, etc. Among these, eucalyptus has been riased on a much larger area (13.49 km^2 in 1978-79 to 31.56 km^2 in 1981-82) than others. Thus, eucalyptus by virtue of being a fast growing exotic species has been given more prominance in government's afforestation programme. As far as the plantation of coniferous species is concerned a rather poor performance is indicated (Table 2.9). It may be either due to the inaccessibility of suitable sites of lack of interest. Among the coniferous species, chir is planted on 84.34 mk^2 followed by deodar (112.97 km^2) upto 1981-82.

Plantation Raised During Plan Periods

In response to the National Forest Policy of 1952, the 'Van Mahotsava' programme was launched to revive the old association of Indian culture with forests, to promote better landuse and to make people conscious of the importance of trees and forests in respect to their daily life and the country's economy as well as to maintain the ecological balance. Keeping this in view, the government has paid increasing attention in each successive plan period towards afforestation. Before taking the area under plantation each state government was asked by the centre to indicate the terrain characteristics and land utilization pattern of the respective states so that forest area may be developed according to the national norm. Here an attempt has been made to evaluate the progress made in plantation during different plan periods under the control of Forest Department.

Table 2.9: **Species-wise plantation in U.P.**

Species	Area planted (in km^2) in 1981-82	Total area planted upto 1981-82
A. Non-coniferous (Broad-leaved species)	225.51	5978.15
1. Teak	39.89	486.39
2. Sal	3.85	85.04
3. Eucalyptus	31.56	889.22
4. Sissoo (Shisham)	5.59	289.52
5. Semal, Gutel and Ailanthus	3.20	102.00
6. Kanju	0.03	20.39
7. Khair	3.55	N.A.
8. Bamboo	4.44	81.76
9. Paper mulberry/mulberry	0.09	20.25
10. Mixed and miscellaneous	133.31	3152.99
B. Coniferous	15.81	513.43
11. Deodar	4.07	112.97
12. Chir	8.64	84.34
13. Cyprus	0.58	8.92 (Including blue pine)
14. Fir and Spruce	0.01	1.90
15. Others	2.51	305.30
Total	241.32	6491.58

N.A. = Not available.

Prior to the commencement of First Five Year Plan, 30,134 ha area was planted. The first Five Year Plan (1951-56) showed great concern towards the rehabilitation of the forest cover which had been earlier over exploited either to fulfil the Second World War requirements or due to indiscriminate cutting of trees by the outgoing zamindars. In this period man made forests were raised over an area of about 24,388 ha. The plan outlay on forestry was Rs.1.38 crores.

In the Second Five Year Plan Period (1956-61) the programme of consolidation of forest area was continued. In this plan more emphasis was laid towards plantation of industrially and commercially viable plant species. To facilitate the forest development programmes, roads and other associated infrastructures were also constructed. During this plan period 76,621 ha land was put under forest which is more than three times the area planted during the First Plan Period. The outlay on the forest (Rs.2.68 crores) was nearly double than that of the previous plan period. Out of the total investment, 40.23 per cent funds were allocated for the plantation of fast growing species, raising of plantation of economic and industrial importance, reclamation of ravines and fuel wood plantation-cum-farm forestry.

In the Third Five Year Plan (1961-66) more emphasis was laid to bridge the gap between the demand and supply of forest produce. The area planted under this plan was 97,891 ha. To avoid wastages, better techniques of timber extraction and of proper seasoning and preservation, were also recommended. In addition, more attention was given to fast growing plant species so that the increasing demand of raw material by the paper and pulp industry could be met. The total plan expenditure on forestry was Rs. 9.45 crores, of which 37.32 per cent (Rs.3.52 crores), was allocated to main plantation schemes of the state.

The Fourth Five Year Plan could not commence in time due to certain constraints before the nation and therefore, its execution was delayed by three years. During this short tenure (1966-69) the total area planted was recorded as 89,653 ha with an investment of Rs. 7.16 crores.

The main objectives of the Fourth Five Year Plan period (1969-74) were - (i) Planting of quick growing species such as eucalyptus, (ii) plantation of economic species like teak, semal, gutel, sissoo, and bamboo, etc., and (iii) to develop plantation for rehabilitation of denuded forest. In addition, it was also proposed to take up wasteland for fuel wood plantation. By the end of Fourth Plan, 186,016 ha land was brought under the above schemes. Out of this, broad-leaved species occupied 136,756 ha area whereas 9,276 ha was planted by the coniferous species. Among broad-leaved species, the area planted to teak was 10,593 ha, semal: 1,657 ha, Sissoo: 882 ha, bamboo: 32,723 ha and eucalyptus: 33,380 ha. Among coniferous species, the maximum area was planted by chir (2,879 ha) and it was closely followed by deodar (2,430 ha). The total capital outlay was of the order of Rs.17.58 crores; of this Rs.10.45 crores was dedicated to main plantation schemes of the state which included reclamation of ravines and others.

The basic objective of the Firth Five Year Plan (1974-79) was to initiate measures for increasing production of industrial wood, to develop the much needed farm forestry and improvement of denuded forests to raise fuel and timber supply mainly in the rural areas. Much more emphasis was given on farm forestry because the trees near the settlements have gradually disappeared in meeting the fuel and other domestic needs. By the end of this plan, a total of 1,21,024 ha land was developed under forestry with a capital outlay of Rs.30.33 crores. Out of this, major share (72.35%) went to the government's main plantation schemes. Of the total area planted, the broad-leaved species covered 1,16,004 ha as against 5,020 ha under coniferous species. Among broad-leaved species, teak was planted over an area of 14,002 ha, sissoo on 5,213 ha, bamboos on 20,620 ha and eucalyptus on 17,999 ha. In coniferous species, deodar and chir were raised over an area of 1,408 and 1,218 ha, respectively.

Owing to a change over in political power, the Sixth Five Year Plan was delayed by one year. During this period (1979-80) 21,125 ha ofland was planted at a cost of Rs.11.34 crores

Table 2.10: **Plantation raised upto 1981-82 on the areas under the control of Forest Department**

Period	Area planted (ha)	Total plan investment (in crores)	Expenditure under main plantation schemes (in crores)	Percentage of column 4 to column 3
1. Prior to First Five Year Plan Period	30,134	-	-	-
2. During First Five Year Plan period	24,388	1.38	0.22	16.05
3. During Second Five Year Plan period	76,621	2.68	1.08	40.37
4. During third Five Year Plan period	97,891	9.45	3.52	37.32
5. During Annual Plan Periods (1966-69)	89,653	7.16	4.46	62.39
6. During Fourth Five Year Plan period	1,46,016	17.58	10.48	59.61
7. During Fifth Five Year Plan period	1,21,024	30.35	21.96	72.35
8. During 1979-80	21,125	11.34	7.72	68.13
9. During Sixth Five Year Plan period	42,306	36.26	30.55	84.26
Total	6,49,158	116.20	79.99	68.84

Sixth Five Year Plan commenced in 1980-81. Due to lack of latest information regarding the plantation raised in the state for the complete plan period which has just ended, we are relying on the statistics of only two consecutive years (1980-82).

During the two years, 42,306 ha area was planted by various species with a total expenditure of Rs. 36.26 crores, of which 84 per cent was earmarked for the main plantation schemes.

From the First Five Year Plan to 1981-82 a total area of 7,64,321 ha has been planted, inclusive of 1,15,163 ha area which is not under the control of Forest Department.

Area Planted in Different Geographical Regions Under Plan Schemes

Plantation of trees was made in different geographical regions considering their suitability as well as the future requirements of forest resources in various sectors of the Indian economy. Looking into the bio-climatic and other governing locality factors, an attempt has been made here to evaluate plantations carried out for specific usage in different geographical regions of the state under various plan schemes.

Of the total area planted under plan schemes, maximum development of man-made forests (44.56%) has occurred in Tarai and Bhabar regions followed by Vindhyan region (24.55%), Indo-Gangetic plains (21.84%) and the hills (10.04 %). The factors in favour of Tarai and Bhabar regions were the location of paper and pulp and straw board industries, and supply of resin to industries in and around this region. To meet the industrial requirements, plantations of species of industrial importance covered an area of 146,978 ha (Table 2.11). Water logging and high incidence of malaria in Tarai region has prompted the plantation of fast growing species over an area of 95,644 ha. It is expected to reduce the magnitude of these two hazards in time.

Table 2.11 indicates that since the beginning of the plan schemes more attention has been given by the forest department towards plantations of species related to economic and industrial importance. So far the areal coverage of plantation is concerned, maximum area (30.83%) was devoted for this purpose. It is closely followed by plantations of fast growing species (28.94%), and reclamation of revines (14.22%). In addition, plantations were also made on degraded forested areas (4.13%) and areas prone to erosive action (44.6%).

Table 2.11: **Area planted under plan schemes by the Forest Department upto 1981-82 (in ha).**

Plantation for various purposes	Geographical regions				Total
	Hills	Tarai & Bhabar	Indo-Gangetic plain	Vindhyan	
Plantation of species of Economic and industrial importance	30,038	1,46,978	13,842	-	1,90,058 (30.83%)
Plantation of Fast Growing Species	-	95,644	22,384	61,106	1,79,134 (28.94%)
Rehabilitation of degraded forests	2,920	776	6,347	15,560	25,603 (4.18%)
Fuelwood Plantation	-	1,562	10,041	-	11,603 (1.87%)
Conversion of low value forest to more valuable forest	-	10,226	3,965	-	14,191 (2.29%)
Ravine reclamation	-	-	46,072	41,993	88,065 (14.22%)
Under soil conservation scheme	6,309	505	8,263	12,527	27,604 (4.47%)
Area planted under non-Plan	22,929	20,127	18,124	20,786	81,966 (13.24%)
Grand Total	62,196 (10.04%)	2,75,858 (44.56%)	1,28,998 (20.84%)	1,51,972 (24.55%)	6,19,024 (100.00%)

THREE

Economic Assessment of Forest

Forest can be viewed as one of the important economic assets for the country's prosperity and a major segment of it is owned by the government. Hence, it may be treated as a public sector undertaking. Generally, the economic worth of a resource undertaking is measured in terms of state exchequer. From the sale of commercial timber and other products like bamboos, katha, khair, resin, thatching grass, honey and wax, gum, canes, medicinal herbs and tendu leaves, the forest contributed Rs. 574.76 million in 1981-82 to the state exchequer. This may be considered as an effective contribution to the state's economy. Nevertheless, Uttar Pradesh produces only one-third of its demand of forest goods and the rest is imported from other neighbouring states. A big state like Uttar Pradesh could do better in this respect. It is hoped that in future, seeing the economic importance of forest resources as well as several indirect advantages, due emphasis will be laid to the development of forest.

Though forestry shares, relatively a small part in the states' economy when compared with agriculture or minerals, the forest industry provides mankind a wide variety of important products. Wood has always been used in abundance, wherever it was available, as fuel both for domestic and industrial purposes. After fuel, the most important use of wood is in the building and constructional industries as sawn timber, plywoods and veneers, and fibre boards. It is also used as a major raw material in the paper and pulp industry. With the growth of population, and development in various sectors, this industry is constantly increasing. This creates an ever increasing demand for timber. Wood cellulose is the basis of the synthetic fibre such as rayon and acetate and is therefore, important in modern textile industries. In addition to these major products, forests supply a wide range of minor products such as gum, resin, honey and canes, etc. With the rapid growth in the demand of timber and timber products, consumption of wood has continually expanded. So care should be taken that every tree cut must by law be replaced by atleast two new trees. This has dual advantage of ensuring timber supply and protecting soil - the most valuable resource of the planet earth. Taking these facts into account, it has been attempted here to review the position of forest under different economic regions, outturn of major and minor forest products as well as their economic value. The avenues for employment in the forestry sector have also been assessed.

Position of Forest under Different Economic Regions

The forest resources of U.P. have to play a very important role in strengthening of the economy of the state. It provides fodder to the animals, wood for fuel and raw materials for varied types of industries. The area under forest in different economic regions has been analyzed below. Economic regions are those regions where homogeneity in general economic landscape is found. The forest department for its convenience has divided the state into five broad economic regions (Fig.10).

Hill Region

It consists of eight districts where forest is spread over an area of about 23,820 km^2. This region poss- esses the maximum forest areal coverage, i.e., 58.54 per cent (Fig. 10 & 11). In this economic region the maximum area (17.09%) under forest is found in Uttar Kashi located in the extreme northern part of the state, followed by Chamoli (9.10%), Nainital (8.86%), Tehri (6.62%) and Garhwal (6.01%), whereas the minimum (3.38%) is found in the Pithoragarh district. In contrast to official forest statistics the interpretation of landsat imageries closely indicate the poverty in vegetal cover in the Indian Central Himalayan regions. Table 3.1 shows that among the hill regions the highest percentage of forested land is obtained in Nainital district i.e., 44.63% followed by Tehri (35.90), Pauri (35.80), and Almora (33.30). In spite of this, no one district fulfil the requisite national norm. Increased human population and expanding agriculture are the principal causes of forest deterioration (Singh et. al., 1985). The census data of 1980-81 showed the population of Kumaun as 23,83,163 individuals. This population is 3.4 times higher than that reported for 1981 (Joshi et.al., 1983). Though the results of landsat imageries are not available either on region or district basis, yet in author's view the situation regarding the forest coverage will be more or less the same.

Central Region

This region is stretched over nine districts and occupies an area of 2363.24 km^2 (5.80%). Among them Kheri has the largest

Table 3.1: **Area (Km^2) under non-forested landuse in different districts on Kumaun and Garhwal Himalaya**

Landuse	Kumaun			Garhwal				
	Almora	Nainital	Pitloragarh	Pauri	Uttarkashi	Tehri	Chamoli	Dehradun
Non-forest								
Snow	339	0	2629	0	1176	366	3431	0
High altitude meadows	228	0	504	0	192	142	774	0
Other non-forest	3014	3754	3351	3492	1670	2327	2799	2382
Total non-forest	3581	3754	6484	3492	3038 +3744*	2835	7004	2382
Forest								
Poor forest	404	872	491	826	293	900	670	253
Medium forest	1138	1900	1343	972	491	358	1260	204
Good forest	238	254	525	150	440	330	69	257
Total forest	1780 (33.20%)	3026 (44.63%)	2359 (26.67%)	1948 (35.80%)	1224 (15.28%)	1588 (35.90%)	1999 (22.20%)	714 (23.06%)
Grand Total	5361	6780	8843	5440	3006	4423	9003	3096

Gap Area (for which images were not available), includes mainly snow and high altitude meadows.

Source: J.S. Singh (ed.), Environmental Regeneration in Himalaya, 1985, p.241.

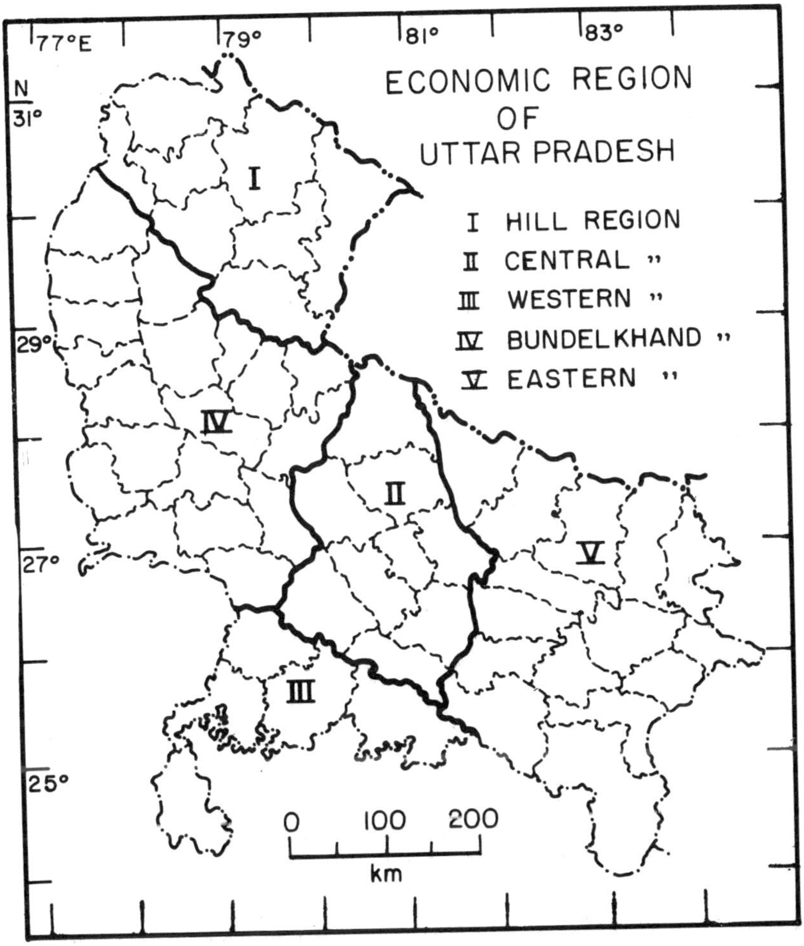

FIG.10. ECONOMIC REGION OF UTTAR PRADESH

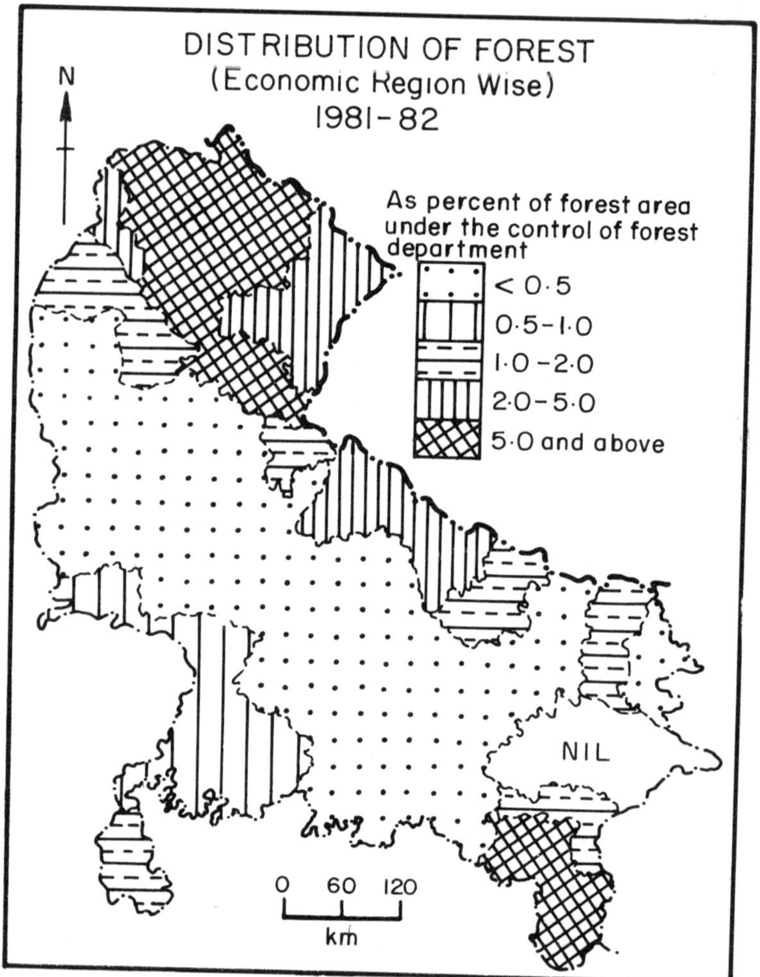

FIG.11. DISTRIBUTION OF FORESTS (ECONOMIC REGION-WISE) 1981-82

area under forest (4.08%), whereas Sitapur has the least (0.10%). In this region, except Kheri, all districts have an ear below 0.4 per cent under the forest (Table 3.2).

Table 3.2: **Forest area in different economic regions (1981-82)**

Economic region/ district	Area under control of forest deptt. (in km^2)	Percentage of the total forest area under the control of forest deptt.
A. Hill Region	23,819.65	58.54
1. Almora	1,494.81	3.67
2. Pithoragarh	1,378.44	3.38
3. Nainital	3,604.78	8.86
4. Garhwal (Pauri)	2,447.65	6.01
5. Chamoli	3,704.87	9.10
6. Tehri-Garhwal	2,694.26	6.62
7. Dehradun	1,539.96	3.78
8. Uttar Kashi	6,954.88	17.09
B. Central Region	2363.24	5.80
9. Barabanki	78.34	0.19
10. Fatehpur	48.64	0.12
11. Hardoi	111.73	0.27
12. Kanpur	118.62	0.29
13. Lakhimpur (Kheri)	1,661.60	4.08
14. Lucknow	94.72	0.23
15. Rae Bareli	48.02	0.12
16. Sitapur	43.28	0.10
17. Unnao	158.29	0.39
C. Western Region	3,833.55	9.42
18. Agra	394.42	0.97
19. Aligarh	8.08	0.02
20. Bijnor	708.02	1.74
21. Badaun	69.03	0.17
22. Bareilly	4.11	0.01
23. Bulandshahar	103.20	0.25
24. Etah	11.66	0.03
25. Etawah	386.83	0.95
26. Farrukhabad	32.70	0.08
27. Mainpuri	65.74	0.16
28. Mathura	15.75	0.04
29. Meerut	79.93	0.19
30. Moradabad	119.21	0.29
31. Rampur	66.11	0.16
32. Pilibhit	785.03	1.93

33. Muzaffarnagar	72.76	0.18
34. Saharanpur	773.15	1.90
35. Sahjahanpur	111.77	0.27
36. Gaziabad	25.55	0.06
D. Bundelkhand Region	2,403.69	5.90
37. Banda	771.81	0.02
38. Hamirpur	373.18	0.92
39. Jalaun	257.31	0.63
40. Jhansi	325.44	0.80
41. Lalitpur	669.95	1.64
E. Eastern Region	8,260.83	20.30
42. Allahabad	201.42	0.49
43. Azamgarh	2.38	-
44. Bahraich	1,017.73	2.50
45. Ballia	-	-
46. Basti	82.34	0.20
47. Deoria	11.51	0.29
48. Faizabad	12.68	0.03
49. Ghazipur	-	-
50. Gonda	718.24	1.76
51. Gorakhpur	556.30	1.36
52. Jaunpur	2.66	-
53. Mirzapur	4,852.62	11.92
54. Pratapgarh	4.44	0.01
55. Sultanpur	24.97	0.01
56. Varanasi	774.04	1.90
Total	40,687.10	100.00

Western Region

Though this region comprises eighteen districts, so far the area under forest is concerned it covers only 3,833.55 km^2 which is 9.42 per cent of the total area. Among the various districts Pilibhit, Saharanpur and Bijnor have. Whereas the remaining districts have less than 1 per cent area under the forest (Fig. 10 & 11).

Bundelkhand Region

It comprises most of the districts of southern upland viz., Banda, Hamirpur, Jhansi, Jalaun and Lalitpur. These five districts together account for 5.90 per cent of the forest area under the control of the Forest Department. Lalitpur has the highest percentage under forest (1.64%), whereas Banda has the least, i.e., 0.02 per cent. Being situated in uplands, these districts have only a nominal area under forest which is much less than the norm recommended by the government. There is a need of much more effort to plant trees in this area so that this region may occupy a sound position in forest resources.

Eastern Region

This region stretches right from Allahabad to Ballia districts. These eastern most districts of the state have the second largest area under forest, after the hill region i.e., about 8,261 km^2. Among the fifteen districts of this region, Mirzapur has the highest forested area (11.92%), whereas Ghazipur, Ballia, Jaunpur and Azamgarh are without forest. Of the remaining districts only Gonda (1.76%), Gorakhpur (1.36%) and Varanasi (1.90%) have more than 1% area under the forest.

It is apparent from Fig.11 that leaving a few districts such as Lakhimpur and Mirzapur, the Central, the Western, the Bundelkhand and the Eastern region exhibit a very poor state of affairs as far as forest resources are concerned. Only the hill region has a sizeable area under forest and is of high economic value. However, the hill forests are located at high altitude and in inhospitable terrain with poor accessibility. Hence, it can be concluded that forestry is in bad shape in Uttar Pradesh, the most populous state of India. The hill region alone cannot fulfil our needs. Therefore, in the national plan to develop the state, forest wealth should get a paramount importance not only from the economic point of view but also due to the other indirect environmental advantages.

Forest Area by Merchantability

Forests which are within the reach of economic management or exploitation as sources of forest products including immature forests and managed forests where fellings are restricted are known as the merchantable forests. And the forest which is not yet managed or exploited owing to inaccessibility or on account of forest produce in them being unsaleable, is known as non-merchantable forsets.

The economic accessibility of forest should be considered in terms of present prices of wood and cost of its extraction. Merchantability of forest depends very much upon the linkage of roads and means of transportation and the nature of terrain. It will be more in area where these conditions are favourable and less in non-accessible areas. From the view point of merchantability about 62 per cent area (2,530,620 ha) is merchantable, 13.94 per cent forest area is potentially merchantable and the remaining area is unmerchantable (Table 3.3). The main use of forest is the extraction of timber and this will likely to remain the most important single use of forest in the coming decades also. In accordance, 64.89 per cent of all forest classes is dedicated to production of timber. Of the total area dedicated to timber, 73.02 per cent area is located in merchantable forest whereas in the potential merchantable and unmerchantable classes 10.43 and 16.55 per cent area, respectively, can be utilized for extraction of timber in near future. Besides timber, however, forests have other economic values, exploitation for which has been increasing in recent years. Of the toal forested area under the control of Forest Department, 35.11 per cent is used for various other purposes. Among this 42.20 per cent comes from merchantable forest. It is notable that in potentially merchantable as well as in unmerchantable forest, a large area is devoted to other usage as compared to area dedicated to timber.

Table 3.3: **Forest area according to merchantability (1981-82)**

Classification/ Status	Area (in ha.)	As percentage of the total forested area under the control of forest Department
1. Merchantable	2,530,620	62.19
(a) Dedicated to timber	1,927,812	47.38
(b) Others	602,808	14.81
2. Potentially merchantable	567,182	13.94
(a) Dedicated to timber	275,426	6.77
(b) Others	291,756	7.17
3. Unmerchantable	970,916	23.86
(a) Dedicated to timber	436,970	10.74
(b) Others	533,946	13.13
Total area dedicated to timber	2,640,208	64.89
Total area under others	1,428,510	35.11
Grand Total	4,068,718	-

Outturn of Major Forest Products

As it has been already discussed, the main use of forest is the extraction of wood because wood has a wide range of usage in construction and as a raw material for many industries but the most important single use of wood is fuel (Fig.12). It can be readily seen from table 3.4 that about 947,112 cu. m. of industrial wood and 1,749.806 cu.m. of fuel wood were annually extracted by the various agencies (departmental, purchaser, right holder, concessionist and free grantees) in 1981-82. The production of these two items was higher in 1978-79 (1,149,414 cu.m. industrial wood and 2,355,274 cu.m. fuel wood). About 65 per cent of the total wood removed from forest all over the state is used as fuel. This large percentage meets the fuel requirement of the entire rural and a part of urban population. Till now a significant proportion of the demand for fuel wood as well as industrial wood is met from sources other than the departmental agencies. The purchaser alone produces 878,750 cu.m. (92.78%) industrial wood and 1,074,942 cu.m. (61.43%) fuel wood. But in the coming decades to meet the needs of rapidly growing population more reliance would have to be placed on the forest department. The above mentioned two forest products provide a sum of Rs. 454.39 million to the state, and of it, timber contributes a total of Rs.434.78 million which is 95.68 per cent of the total value of major forest products (Table 3.4).

Timber is used in three main ways: (a) Sawn wood: Currently with changing taste and rise in income, sawn wood is actually replacing round wood. Earlier the round wood was utilized mainly for the construction of thatched huts and mud houses. But with the development in technology, sawn wood is being used in various functions both in rural and urban areas in increasing quantum. This would lead to a large economy in the use of wood. Besides there is a large demand for sawn wood in heavy-duty construction works such as in the form of railway sleepers. It is also used for plank, boards, beams and so forth in house building. In 1981-82 the actual out turn of timber was 149,797 cu. m. Right from 1978-79 to 1981-82 the out turn of sawn wood has decreased.

(b) Pulp wood, match-wood, etc.: Timber is also extensively used in match and pulp and paper industry, sports goods industry, agricultural equipments and wooden handicrafts. In 1978-79, the outturn of timber in terms of pulp wood, match wood, round wood, etc. was 1,149,354 cu. m. which decreased to 947.102 cu. m. in 1981-82. It is notable here that there is no definite trend in the production of timber for pulp and paper wood (Table 3.5).

(c) Outturn of timber in terms of logs, poles etc.: In 1981-82, out of the total timber produced, 495,153 cu.m. were used in the form of rough logs, 149,797 cu. m. as sawn logs, 158,376 cu. m. as rough poles and the remaining for miscellaneous purposes.

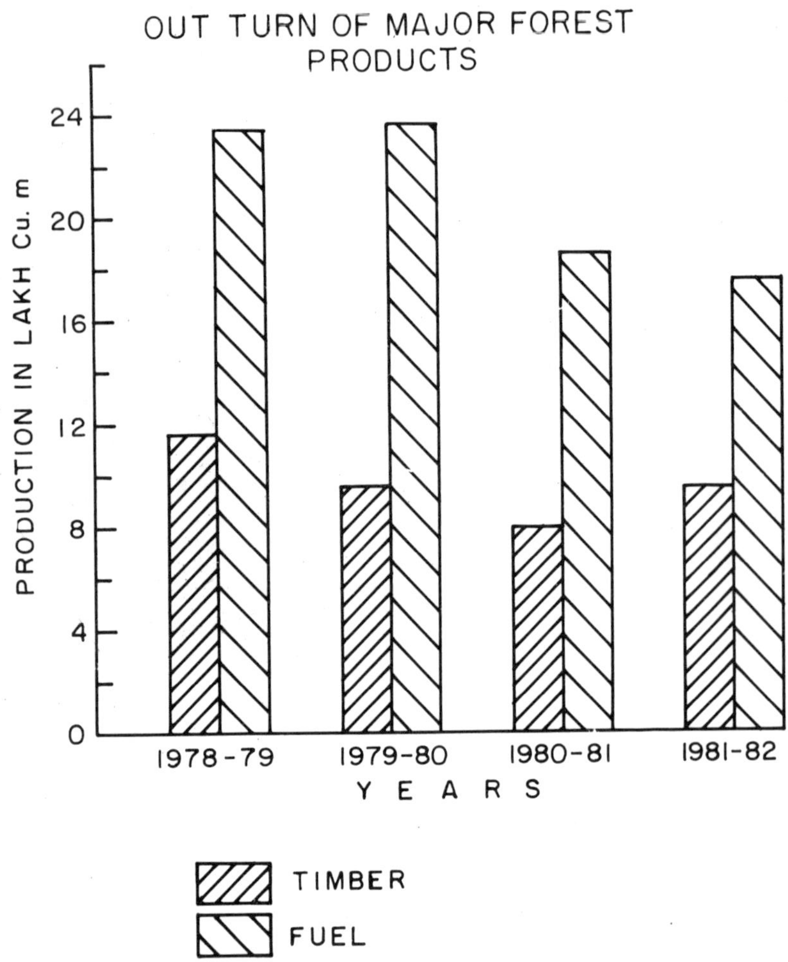

FIG.12. OUT PLAN OF MAJOR FOREST PRODUCTS

Table 3.4: **Out turn of major forest products (in cu. m.)**

Item	1981-82		1980-81		1979-80		1978-79	
	Production	value (in ,000 Rs.)	Production	value (in ,000 Rs.)	Production	value (in ,000 Rs.)	Production	value (in ,000 Rs.)
Timber	947,112	434,785	796,215	356,358	961,987	327,770	1,149,414	361,946
Fuel	1,749,806	19,609	1,861,767	17,100	2,359,103	16,514	2,355,274	14,241
Total	2,696,918	454,394	2,657,982	373,458	3,21,090	344,284	3,504,688	376,137

Table 3.5: **Out turn of timber in terms of sawn wood and pulpwood etc. (in cu.m.)**

Timber	1981-82	1980-81	1979-80	1978-79
Sawn wood	149,787	187,795	195,455	245,223
Round wood, pulp-wood, match wood etc.	797,315	608,420	766,532	904,131
Total	947,102	796,215	961,987	114,9354

The production of rough logs has been high at all times (Table 3.6).

Table 3.6: **Out turn of timber in terms of logs, poles etc.(in cu.m.)**

Timber	1981-82	1980-81	1979-80	1978-79
Rough logs	495,153	446,423	539,516	615,025
Rough poles	158,376	75,077	132,636	90,902
Sawn logs	149,797	187,795	195,455	245,223
Miscellaneous	143,786	86,920	94,380	198,264
Total	947,112	796,215	961,987	1,149,414

Out turn of timber by species

On the basis of species, the forests of U.P. can be categorized into two broad divisions: (a) non-coniferous and (b) coniferous. The non-coniferous species have always yielded more timber than the coniferous species. This is because of more areal coverage of non-coniferous species. In addition, non-coniferous forest has a greater variability in species composition, compared to coniferous forest. Non-coniferous species are generally found in all bio-climatic regions whereas the coniferous species flourish only in the mountainous and sub-mountainous areas. Table 3.7 gives species-wise out turn of timber in 1981-82. In 1981-82 the out turn of timber extracted from various species was 947,112 cu.m., out of which 547,034 cu. m., (60.61%) was obtained from non-coniferous species. Among all the non-coniferous species, the maximum out turn of timber has been obtained from Sal (193,669 cu. m., i.e., 33.2%)

followed by Eucalyptus (19.98%). These two have jointly yielded 53.72 per cent of the total timber extracted from the forests of U.P. Among coniferous species, the maximum out turn of timber was gained from chir, i.e., 337,253 cu. m. which is 90.38 per cent of the total production of timber from coniferous species.

Table 3.7: **Out turn of timber by species (cu.m.), 1981-82**

Non-conferous			Coniferous		
S.No.	Species	Production	S.No.	Species	Production
1.	Sal	193,669	1.	Chir	337,253
2.	Teak	9,627	2.	Deodar	12,465
3.	Asna	9,187	3.	Fir & Spruce	14,155
4.	Shisham	15,645	4.	Kail	9,089
5.	Semal	2,258	5.	Oak	-
6.	Gutel	578	6.	Cyprus	116
7.	Haldu	4,543			
8.	Jamun	5,897			
9.	Tun	765			
10.	Khair	3,535			
11.	Kanju	2,167			
12.	Dhau	-			
13.	Siris	606			
14.	Jhinger	1,742			
15.	Kharpat	457			
16.	Sandan	1,770			
17.	Eucalyptus	114,692			
18.	Miscellaneous	206,597			
Total		574,034		Total	373,078

Out of the total timber yield, the Forest Corporation of U.P. has extracted 479,000 cu. m. which is nearly half (50.45%) of the timber produced in 1981-82. It was only 25,000 cu. m. in 1974-75. During 1974-75, the maximum out turn ot timber was obtained from Sal (23,000 cu. m.) but now the position has completely changed. At present the maximum timber extracted by Forest Corporation of U.P. is from chir (246,000 cu. m.) followed by Sal and Eucalyptus (114,000 and 86,000 cu. m. respectively). The production of timber from sal was higher till 1977-78 but after this the production of chir took the lead and right from 1978-79 to 1981-82 it was nearly two times more than the producation of timber from sal. On the whole, the out turn of timber from non-coniferous species was more than from the coniferous species.

Out turn of Minor Forest Products and their Value

Forests yield a wide range of minor products such as Bamboo, katha, resin, tendu leaves etc. These are used in various industries as raw materials and contribute about Rs.135.50 million to the state exchequer. Table 3.8 gives an account of the out turn of minor forest products and their value. Among the various minor forest products, resin contribute the maximum amount (Rs. 58.90 million) to the forest department followed by bamboo (Rs. 9.61 million), khair (Rs. 12.42 million), baib grass (Rs. 3.15 million), tendu leaves (Rs. 23.03 million), thatching grass (Rs. 1.96 million) and katha (Rs. 2.05 million) etc. (Fig. 13). Resin and tendu leaves (used for biri making) share 43.45 and 16.99 per cent respectively of the total value of minor forest products. Resin is derived from the resinous material exuded from coniferous species particularly chir, which is collected or systematically tapped and processed for a variety of uses. It is widely used in chemical industry. Turpentine - the distilled form of resin is used in paints. It is mostly supplied to the I.T.R. company Bareilly and co-operative societies, etc. It is evident from the Table 3.9 that forest contributes Rs.589.9 million from the major and minor forest products. About 77 per cent of the total value of forest produce in the state comes from major forest produce and the remaining from the minor forest produce. Roughly the same trend persists right from 1966-67 onwards. In 1966-67 the share of major forest produce was 69.84 per cent while was 66.48 per cent of the total forest produce of the state in 1970-71 (Table 3.9).

From 1930-31 to 1974-75 the production of resin witnessed a significant upward trend (Fig.14). In 1930-31 the out turn of resin was 30,596 qt. which rose to 54,536 qt. in 1950-51. Further, the production became nearly three times (140,816 qt.) in 1960-61 as compared to 1950-51. The production of resin increased till 1974-75. But after that there is a marked declining trend in the production of resin. The year 1974-75 recorded the highest production of resin, i.e., 417,926 qt. valued at Rs. 74.80 million.

Tendu leaves, the second important product of the minor forest noticed an upward trend right from 1960-61 to 1970-71. After this period the production of tendu leaves registered a fluctuating trend. During the last 26 years the production of tendu leaves was highest in 1978-79. It was 235,060 qt. valued at Rs. 2489 million.

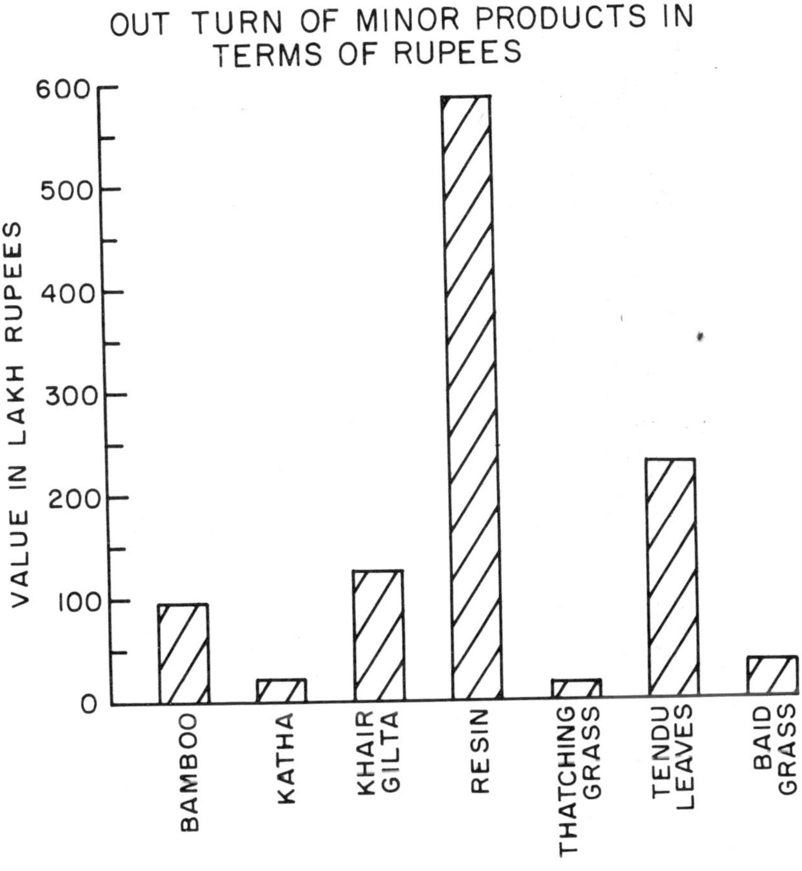

FIG.13. OUT TURN OF MINOR FOREST PRODUCTS IN TERMS OF RUPEES 1981-82

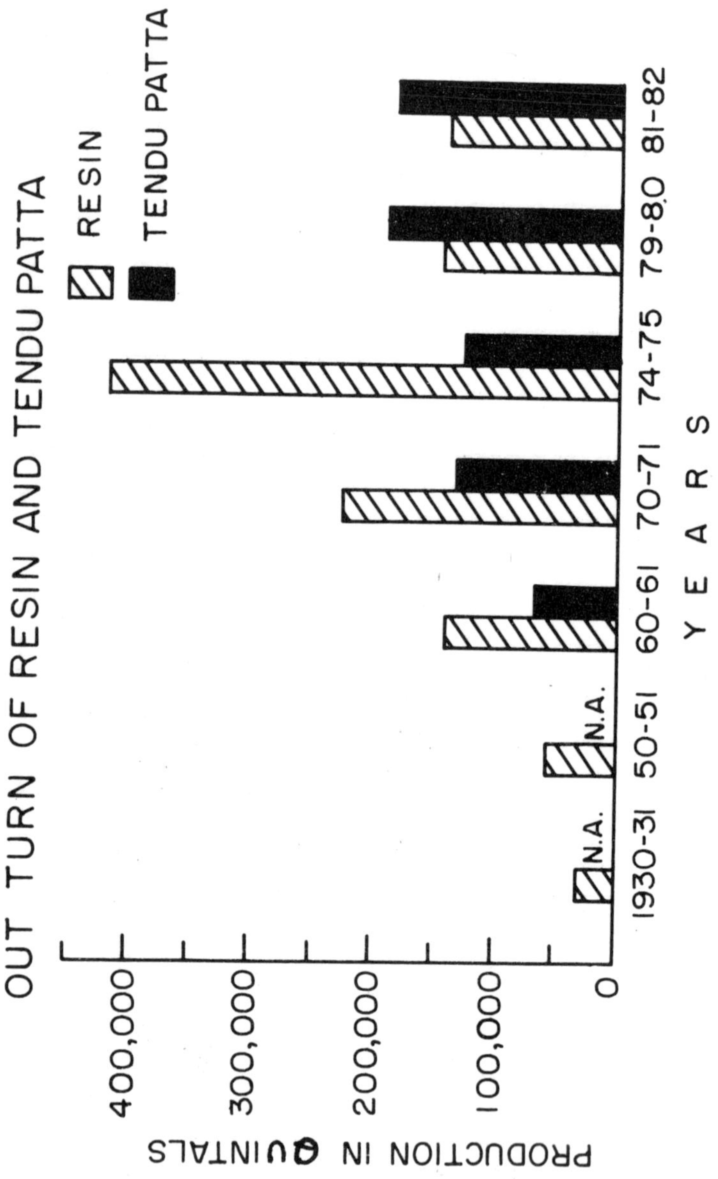

FIG.14. OUT TURN OF RESIN AND TENDU PATTA

Table 3.8: **Out turn of minor forest products and their economic value 1981-82**

Item	Unit of quantity	Quantity	Value in lakh Rs.
Bamboo	Nos.	12,276,529	96.15
Katha	Quintals	411	20.55
Khair gitta	Quintals	33,769	124.19
Resin	Quintals	141,241	588.82
Thatching grass	Quintals	179,460	19.68
Honey wax	Quintals	453	5.24
Gum	Quintals	1,290	3.74
Canes	Qtls.	429	-
	Bundles	22,716	5.13
Medical herbs	Quintals	775	1.33
Tendu leaves	Quintals	184,180	230.37
Baid grass	Quintals	89,866	31.54
Total		-	1,355.21

Out turn of Major Forest Produce in Hills

Hills are considered the treasure of forest products but in the wake of development, man due to his greed and need, has expoiled as well as despoiled forests considering it as an inexhaustible resource. Today (1981-82) the hills contribute about 3/4th (72.11%) of the total forest produce. To the total production of sawn and fuel wood in the state, the hills contributed 62.15 and 78.20 per cent, respectively. Statistics clearly indicate that the hills are not as important in the production of timber as in fuel. It is in the fitness of things because fuel is the major necessity of hillmen to protect themselves from the vagaries of severe cold. Table 3.10 does not show any definite trend of the out turn of timber and fuel. For example, the production of timber and fuel wood was 595,115 cu.m. and 1,548,230 cu. m. respectively in 1971-72 as against 400,762 cu. m., and 1,368,436 cu. m. in 1981-82.

Out turn of some important species of timber (in terms of sawn wood)

Here the out turn of timber from the twelve important species* has been considered. Table 3.11 does not show any

---

* Sal, teak, sissoo, khair, chandan, ashna, kanju, haldu (non-coniferous); kail, deodar, fir and spruce, chir. (coniferous)

Table 3.9: **Comparative value of forest produce (in, 000 Rs.)**

Year	Miner products		Major products			
		As percentage of total	Fuel	Timber	As percentage of total	Total value
1966-67	49,588	31.16	13,133	96,396	69.84	159,117
1970-71	73,448	33.52	14,883	130,768	66.48	219,099
1974-75	131,157	42.25	14,252	178,046	57.75	310,396
1978-79	101,789	21.29	14,241	361,946	78.71	477,976
1981-82	135,421	22.97	19,609	434,785	77.03	589,915

Table 3.10: **Outturn of major forest produce in hills (in cu.m.)**

Year	Timber sawn in hills	Timber sawn in U.P.	Percentage of total timber sawn in U.P.	Fuel in hills	Fuel production in U.P.	Percentage of total fuel production in U.P.
1971-72	595,115	1,042,651	57.07	1,548,230	2,315,275	66.87
1977-78	530,557	858,745	61.78	1,831,745	2,456,553	74.56
1981-82	400,762	655,365	61.15	1,368,436	1,749,806	78.20

definite trend of timber production from the non-coniferous as well as from coniferous species. The production of timber from sal, ashna fir and spruce, haldu, khair and chir has increased from 1930-31 to 1970-71 but after that it has decreased (Table 3.11). Among various species only teak after a short downfall in 1960-61, has shown an upward trend. The production of timber from deodar has increased from 2,917 cu.m. (1930-31) to 29,971 cu. m. (1975-76) but in 1980-81 the production came down to 5,235 cu. m. and increased again to 10,583 cu. m. in 1981-82. It seems that in the extraction of timber it has not been cared which specie is more vital for production of timber.

Per Capita Production of Timber and Fuel during 1981-82

To know the economic potentiality of forest of the state, an attempt has been made to analyse the production of timber and fuel in different economic regions. The per capita production of timber and fuelwood may be taken as an indicator to gauge the potentialities as well as the nature of the area, i.e., which one is in surplus and which one is in deficit. It is evident from table 3.12 that most of the timber and fuel are obtained from the hill districts of the state. In 1981-82, about 400,762 cu. m. of timber and 1,368,436 cu. m. fuel wood were obtained from the hill districts alone. Among the hill districts, maximum out turn of timber (157,067 cu. m.) as well as fuel (589,130 cu. m.) was received from the Nainital while the minimum was obtained from the Tehri district in respect of timber (22,561 cu. m.) and from Uttar Kashi in respect of fuelwood (895 cu. m.). Since maximum production of wood is derived from hill region, so on account of low population, the per capita production of timber (0.0829 cu. m.) as well as fuel (0.2830 cu. m.) is higher than the other economic regions of the state. Among the hill districts, the maximum per capita production of timber is obtained from Uttar Kashi (0.2347 cu. m.) followed by Chamoli (0.1478 cu. m.) and Nainital (0.1381 cu. m.) while the maximum per capita out turn of fuel is obtained from Chamoli (1.494 cu. m.) followed by Nainital (0.518 cu. m.).

In the remaining economic regions, Western region provides about 17 per cent timber while Eastern region has a share of 15 per cent. The Central and the Bundelkhand region contributes 2.41 and 4.25 per cent timber of the total production. In the case of fuel production the Eastern region stands next to the Hill region, having 8.52 per cent share. It is followed by Western and Bundelkhand regions which claim 6.57 and 6.12 per cent share, respectively. Only the Central regions deficient in this regard. It produces only 0.57 per cent of the total fuel production of the state.

Table 3.11: **Comparative outturn of some important species of timber (in terms of sawn wood)**

Species	Production in cu.m.						
	1930-31	1950-51	1960-61	1970-71	1975-76	1980-81	1981-82
Sal	111,654	120,545	191,611	354,860	286,047	165,413	138,112
Teak	-	453	340	705	1,639	7,281	6,554
Sissoo	3,851	2,463	284	10,359	23,167	13,398	11,372
Khair	-	127	4,191	10,666	14,046	3,209	2,480
Sandan	3,087	708	764	1,755	1,215	899	884
Sain (Ashna)	6,966	11,468	12,912	23,861	11,827	8,261	6,656
Haldu	2,692	3,568	5,097	12,474	8,173	1,887	3,137
Kanju	-	2,038	4,587	1,690	1,801	1,705	1,628
Deodar	2,917	9,542	9,471	11,612	29,971	5,235	10,483
Fir and spruce	-	10,987	39,502	39,766	71,493	14,757	11,729
Kail	14,191	5,135	5,380	5,166	10,896	6,502	8,214
Chir	31,885	87,415	107,435	200,030	295,745	136,879	228,392

There are many districts which completely lag behind in timber and fuel production. Agra, Badaun, Mathura (Western region); Kanpur, Fatehpur and Hardoi (Central region); Hamirpur

Table 3.12: **Per capita forest area and per capita production of timber and fuel during 1981-82**

District Economic region	Production (in cu.m.) Timber	Fuel	Per capita Timber cu.m.	Fuel cu.m.
Almora	36,349	111,462	0.0480	0.1472
Uttar Kashi	44,820	895	0.2347	0.0047
Garhwal (Pauri)	25,871	65,695	0.0406	0.1030
Chamoli	53,814	544,057	0.1478	1.4947
Dehradun	15,363	11,780	0.0202	0.0155
Nainital	157,067	589,130	0.1381	0.5181
Pithoragarh	44,917	37,833	0.0919	0.0774
Tehri-Garhwal	22,561	7,584	0.0453	0.0152
Hill region	400,762	136,846	0.0829	0.2830
Aligarh	4	-	-	-
Agra	-	2,690	-	0.0009
Etawah	146	-	0.0001	-
Etah	5	251	-	0.0001
Pilibhit	33,591	10,390	0.0333	0.0103
Farrukhabad	12	-	-	-
Badaun	-	-	-	-
Bareilly	83	1,264	-	0.0060
Bijnour	33,017	36,829	0.0170	0.0190
Bulandshahr	282	-	0.0001	-
Mathura	-	523	-	0.0003
Muzaffarnagar	494	903	0.0002	0.0004
Moradabad	37,749	23,231	0.0120	0.0074
Meerut	184	3,692	0.0001	0.0013
Mainpuri	11	7,772	-	0.0004
Rampur	3,631	9,599	0.0031	0.0081
Shahjahanpur	2,911	1,772	0.0018	0.0011
Saharanpur	253	21,954	0.0001	0.0082
Gaziabad	59	1,172	-	0.0006
Western region	112,432	115,042	0.0029	0.0029
Unnao	7	659	-	0.0004
Kanpur	-	-	-	-
Fatehpur	-	-	-	-
Rae Bareli	7	425	-	0.0002

Lucknow	123	958	0.0001	0.0005
Lakhimpur (Kheri)	15,469	7,185	0.0079	0.0037
Sitapur	127	-	0.0001	-
Barabanki	102	792	0.0001	0.0004
Hardoi	-	-	-	-
Central region	15,835	10,019	0.0008	0.0005
Jalaun	3	637	-	0.0006
Jhansi	357	9,548	0.0003	0.0084
Lalitpur	734	19,656	0.0013	0.0340
Banda	797	77,297	0.0175	0.0504
Hamirpur	-	-	-	-
Bundelkhand region	27,891	107,138	0.0051	0.0197
Azamgarh	-	-	-	-
Allahabad	-	439	-	0.0001
Ghazipur	-	-	-	-
Gonda	33,312	17,451	0.0118	0.0062
Gorakhpur	46,428	68,633	0.0122	0.0181
Deoria	-	788	-	0.0002
Pratapgarh	-	-	-	-
Faizabad	-	1,103	-	0.0005
Ballia	-	-	-	-
Basti	91	143	-	-
Bahraich	12,944	32,471	0.0058	0.0147
Mirzapur	5,604	26,734	0.0028	0.0131
Varanasi	-	1,157	-	0.0003
Sultanpur	26	258	-	0.0001
Jaunpur	40	-	-	-
Eastern region	98,445	149,171	0.0024	0.0036
Grand Total	655,365	1,749,806	0.0059	0.0158

(Bundelkhand); Azamgarh, Allahabad, Ghazipur, Deoria, Pratapgarh, Faizabad, Bailia (Eastern region) do not produce any amount of timber. In the matter of fuel production, Aligarh, Agra, Farrukhabad, Badaun, Bulandshahar (Western region); Kanpur, Fatehpur, Sitapur (Central region); Hamirpur (Bundelkhand), and Azamgarh, Ghazipur, Ballia, Pratapgarh and Jaunpur (Eastern region) show a dismal picture. Thus the situation is worst in the Eastern region.

In the matter of per capita timber and fuel production, most of the districts of these regions are at the low ebb. It has been found that most of the districts of the state on the one hand have the high population and on the other they possess a nominal forested area. Hence, it is natural that the per capita production of fuel and timbr will also be low. The situlation is so acute that barring a few districts such as Mirzapur and Varanasi (especially Chakia tehsil), the villagers use dried cow-dung as fuel. In these areas most of the forest land has been brought either under cultivation or habitation. Whatever the trees are in and around the villages are also disappearing day by day. Because of this, fuel requirement is met from this valuable resource, i.e., the cowdung.

Wood as Raw Material for Industrial Use

With the development of economy old uses of wood pave way to new uses. With the invention of fossil fuel such as coal, petroleum and natural gas, the use of wood as fuel was expected to decline considerably in temporal frame. But with the increasing demand in various sectors of the economy as well as being the raw materials of many industries, the demand of wood has otherwise increased. Hence, the government is making vigorous attempts for raising plantations of suitable species looking to the demand of the various industries. The huge investment on plantations after fifties also justifies the varied use of wood as a raw material for industrial purposes. The various forest produces are being used in the following major industries:

Sleepers supplied to railways

A bulk of wood is supplied to the railways which is one of the biggest industrial enterprises of the country. It consumes a large quantity of wood as sleepers, coak making etc. In 1980-81, 35,609 cu. m. wood was supplied to railways for sleepers (Table 3.13). It amounted to Rs.43.95 millions. Of the total volume supplied to railways 19,600 cu. m. was supplied for broad guage; 10,449 cu. m. for metre guage, 469 cu. m. for narrow gauge and the remaining for special gauge. For the sleepers mostly sal and chir are utilized. In the same year, the total volume contributed by sal for sleepers was 23,110 cu. m. and by chir 12,499 cu. m.

Resin supplied to IIR Company, Bereilly and others

Resin is mostly used in paint making and other chemical industries. The IIR Company located at Bareilly is the main consumer of resin produced in the state. In 1981-82 resin was produced to the tune of 138,287 qt., of which 86,065 qt. (62.23%) was supplied to IIR Company, Bareilly; 354.75 qt. (25.65%) to small scale industries and the remaining to co-operative societies and to others (Table 3.14). The supply of resin fluctuates with demand. In 1978-79, 125,456 qt. resin was supplied to various agencies whereas in 1980-81 it touched the highest level of supply.

Timber supply to Star Paper Mill, Saharanpur

It is the biggest paper mill in the state which has an installed capacity of 30,000 tons of paper annually. Wood is one of the important ingredients of paper industry. The present supply of bamboo is insufficient to meet the large demands of paper mills. Hence, the long fibred coniferous species especially chir, are utilized to augment the supplies. In 1977-78, 15,589 tons twisted chir was supplied to the mill. Its requirement was further increased to the order of 20,706 metric tons in 1981-82. In addition, there are many small paper mills located at Allahabad, Lucknow, Moradabad etc., which also utilize bamboo and other species having medium and long fibre. To meet the future requirements it is advisable to plant long fibred coniferous species instead of bamboo which has medium length fibre.

Timber supplies to Director General of Supplies and distribution (DGS & D)

In 1978-79, the timber of various species valued at about Rs. 1.88 million was supplied to DGS & D. Table 3.15 gives a detailed account of it.

Timber given to various industries

In addition to the above mentioned uses, the timber is widely used in various other industries such as matchwood, plywood, bobbin, sports, pencil, packing cases etc. For each industry the Forest Department has alloted a certain requisite species as well as number of trees for their consumption (Table 3.16). For example, for the matchwood industry - 671 trees of semal, 548 of Gutel and 695 trees of Ailanthus were alloted in 1981-82. Smiilarly, for plywood jhingan, kanju, kharpat, godala, sris, faldu and bahera are utilized. Totally for all the industries referred to above 15,286 trees were alloted in 1981-82 whereas this number was twice in 1979-80 (39,120 trees) and in 1980-81 (33,830 trees). From Table 3.16, it is evident that trees of the coniferous species are less utilized probably because of the soft nature of the wood.

Grazing: A Source of Income

In hilly and plateau regions due to small holdings and lack

Table 3.13: **Sleepers supplied to railways 1980–81**

Species	Broad gauge		Metre gauge		Narrow gauge		Special gauge		Total volume in cu. m.	Total value in Rs.
	Quantity (in nos.)	Volume (cu. m.)	Quantity (in nos.)	Volume (cu. m.)	Quantity (in nos.)	Volume (cu. m.)	Quantity (in nos.)	volume (cu. m.)		
Sal	116,781	10,437	171,818	7,113	15,030	469	NA	5,091	23,110	33,066,110
Chir	102,524	9,163	80,582	3,336	-	-	-	-	12,499	10,891,931
Total	219,305	19,600	252,400	10,449	15,030	469	NA	5,091	35,609	43,957,941

Table 3.14: **Resin supplied to IIR Co. Bareilly and others**

Year	Quantity of resin in Qt.		Others	Small scale industries	Total
	IIR Co. Bareilly	Co-operative societies			
1981-82	86,065	5,437	11,310	35,475	138,287
1980-81	125,784	6,875	4,751	37,206	174,616
1979-80	91,939	11,913	848	29,187	133,097
1978-79	94,798	10,985	2,744	16,959	125,456

Table 3.15: **Timber supplied to Director General of Supplies and Distribution (DGS & D) 1978-79**

Species	Quantity in cu. m.	Value in Rs.
Sal	148	177,823
Deodar	134	151,262
Chir	901	566,556
Fir	1,144	849,448
Others	226	134,644
Total	2,553	1,879,733

Table 3.16: **Timber given to various industries 1981-82**

Name of industry	Species allotted	No. of trees allotted 1981-82	1980-81
Match wood	Semel	671	955
	Gutel	548	1,822
	Ailanthus	695	1,360
Total		1,914	4,137
Plywood	Jhingan	2,311	3,322
	Kanju	2,523	4,734
	Kharpat	127	1,573
	Godala	-	14
	Siris	26	243
	Faldu	26	1,013
	Bahera	115	297
Total		6,182	11,182
Sports	Haldu	33	-
	Pula	246	445
	Gutel	501	499
	Mulberry	12	-
	Faldu	18	43
	Kanju	110	276
Total		920	1,263
Pencil	Godala	7	42
	Baurang	159	331
	Pangar	-	-

	Haldu	-	65
Total		166	438
Packing cases	Dhak	558	2,201
	Salai	-	-
	Gular	152	241
	Gojina	13	29
	Bar Pipal	94	209
	Haldu	1,102	655
	Faldu	1,233	519
	Siris	6	-
	Paper Mulberry	801	853
	Kanju	-	-
	Bahera	167	318
	Babali	1,978	156
	Miscellaneous species	-	-
Total		6,104	5,181
Bobbin	Chambhanik	-	1,225
Miscellaneous	Chir	-	9,560
	Moru	-	12*
	Tun	-	-
	Deodar	-	65*
	Kail	-	30*
	Chambharik	-	338*
	Saur	-	-
	Saur	-	-
	Utis	-	-
	Walnut	-	14*
	Mample	-	-
	Miscellaneous	-	385*
Total		N.A.	10,404
Total		15,286	33,730

* Allotment made to various industries by the Soft Wood Screeing Committee formed for hills.

of irrigational facilities, and severity of climate, agriculture cannot provide sufficient fodder for livestock. Hence, most of the fodder requirements are met from forest areas. It is assumed that on an average a cultivator's requirement is of two plough bullocks and two milk cows and a calf, as the irreducible minimum. To feed these five animals adequately, atleast two acres of grass lands for each animal or ten acres of grass lands for all of them are required. Actually, no household possesses ten acres of grass lands anywhere in the country. This highlights the grave situation in these areas. Hence, people of these areas use natural vegetation as fodder for their animals and this practice continues throughout the year. On one side it destroys the growth as well as the capacity of regeneration and on the other it provides a substantial income to the government. Animals grazed in forests are cows and bullocks, buffalows, goats and sheeps, camels, horses and others. From the statistics it is evident that the area under grazing is about 81.2 per cent of the total forest area under the control of Forest Department and of it about 97.41 per cent area remains open throughout the year for grazing. In terms of animal population the pressure on grazing land is not high as the whole region records 1.89 ha grazing land per animal. From the grazing Rs. 2,946,413 was expected in 1981-82 by the government but grazing fee actually collected was only Rs.480,783. Every year this situation prevails and the government collects only about Rs. three to five lakhs as grazing fee annually. The government loses about Rs.25 lakhs every year, if in it the indirect loss by over-grazing like soil erosion and degeneration etc., is added then the magnitude of the problem is found to be very sever. Hence, government should ban grazing, because with the constant scuffling of the animals feet, the roots get torn out and pulled up and the ground is left bare which ultimately leads to soil erosion.

Employment Opportunities

So far we have discussed many advantages of forest in terms of forest produce and environmental quality; but in terms of employment opportunities the contribution of forest is limited. In labour surplus countries like ours, the investment priorities ought to be decided in terms of employment potential of different schemes, even though they contribute a low proportion to gross national product, national economy. Though forests contribute very little to state exchequer, yet forestry by itself can provide large employment both directly by engaging persons in forests and indirectly in many forest based industries. It provides employment opportunities through management of forests, in lumbering, logging, plantation, collection and processing of various minor forest produce, sawn wood, transportation and social forestry schemes. Skilled to semi-skilled persons are needed to manage forest. These persons supervise the conservation of forest through controlling felling operations and regulating landuse. In man-made forest, labour is required to plant and nourish the selected species. Forest roads are constructed to preserve and make possible efficient use of wood. In the absence of forest roads mature wood might rot and dry trees even may present a fire hazard. In conservation forestry selective felling is practised hence it involves more human hand than the man-made forest. Leaving aside the felling operations, the employment avenues in man-made forest are more than in conservation forestry.

People of the rural areas are more benefitted from the government's plantation and social forestry schemes. Total regular employment in Forest Department of U.P. was 15,510 as on 1.1.83. In addition, thousands of persons are involved in various afforestation and other schemes of the Forest Department on daily wages. In the present circumstances when the employment crisis has widened manifold, the Forest Department can provide jobs to many hands. With the implementation of various developmental schemes of forests, the government can lessen the problem of unemployment on the one hand and it can save the environment from severe environmental degradations caused by various developmental activities on the other.

Since man-made forest provides large employment opportunities besides contributing to the economic wealth of the community, area under man-made forest which is less than one per cent of the total forest area of the state in 1981-82 should be increased to achieve the target laid down by National Forest Policy of 1952.

Contribution of Forest to the State Economy

Table 3.17 sets out the percentage distribution of total state income by different industrial origin at current prices as well as at 1970-71 prices. It is evident that forestry and logging contribute just only 1.1 per cent at current prices. Compared to other spheres of state economy, the contribution of forestry is very low.

In 1980-81 the Forest Department generated a total revenue to the tune of Rs.574.76 million, of which Rs.363.90 million were spent on various non-plan, plan and development activities*. In this way the forests provide Rs. 211.37 million to the state economy (Table 3.18). In the same year the hill region alone contributed 77.30 per cent (Rs. 163.40 million) through

---

* The above figures do not include expenditure incurred under Drought Prone Area Programme' (Rs. 9.2 million) and "Brihad Vriksharopan Schemes'.

Table 3.17: **Percentage Distribution of total state income by industrial origin**

Year	Agriculture and animal husbandry	Forestry and logging	Primary sub-sector	Manufact-uring	All secondary sub-sector	Other sub-sector
I. 1981-82 (at current prices)	45.1	1.1	48.7	12.6	19.9	33.4
II. 1970-71 (at prices)	52.9	0.9	54.2	11.4	19.5	26.3

*Quick estimate
Source: Statistical Diary U.P., 1982 and State Income Estimate Bulletin No.194.

Table 3.18: **Forest revenue and expenditure of U.P.**

Year	Revenue		Expenditure		Income in lakh rupees
	Total forest revenue (in lakh Rs.)	Per sq. km (Rs.)	Total forest expenditure (in lakh Rs.)	Per sq. km. (Rs.)	
1978-79	4,226.07	10,356.60	2,049.74	5,023.19	2,176.33
1979-80	4,517.91	11,131.92	2,412.82	5,945.08	2,105.09
1980-81	4,912.94	12,077.04	3,164.78	7,779.70	1,748.16
1981-82	5,747.62	14,126.37	3,633.90	8,931.31	2,113.72

forest to the state exchequer (Table 3.19). If we consider the forest area under the control of Forest Department on the one hand and the total revenue and expenditure on the other, then we obtain the contribution from the per sq. km forest area to the order of Rs. 5,195 (revenue: Rs. 14,126; expenditure: Rs. 8,931). Table 3.18 gives detailed statistics about the revenue and expenditure from forest during 1978-79 to 1981-82. It is seen that each year the revenue from forest is more than the expenditure. Though revenue as well as expenditure reveal an upward trend, the revenue could not have the same pace as expenditure (Fig.15). During the period 1978-79 to 1981-82 the revenue earned has increased by 36 per cent while the expenditure has increased by 77.28 per cent. It reflects the government's keen interest towards the development of forest and afforestation.

Table 3.19: **Forest revenue and expenditure of hill region (in lakh Rs.)**

Year	Revenue	Expenditure	Income
1978-79	2,020	1,081	939
1979-80	2,337	1,233	1,104
1980-81	2,614	1,423	1,191
1981-82	3,281	1,647	1,634

In terms of expenditure and revenue per ha from forestry we are well below the level achieved by many countries of the world. It is comparable to our traditional method of agriculture where we spend less and earn less. Following the greeen revolution in agriculture we should consider forest also as a crop, and should try to increase the inputs to generate higher outputs. Our forests are rich and have a number of species suited to different bio-climatic regions. Hence, the prospects for future development in these items are promising. The first can provide huge revenues besides large scale employment opportunities, environmental protection and other benefits.

FOUR

Deforestation: A Threat to Environment

Plans for rapid and one sided strategic development are destroying the life supporting environmental opportunities through tremendous pressures on ecological contours. The most frustrating problem of modern times is the craze of over development and speedy exploitation of nature and certainly with the march of new technology - the severity and magnitude of exploitation are making tasks harder and harder each time. The exploding population is another aspect of the massive attacks on ecosystems. There are profound differences in 'the then' and 'the now' ecological aura simply because of the changing attitude of 'gain more and more' is becoming more close to the idea of 'lose more and more'. Now these two have become synonym of one another. Forest is one of the burning example of such conflicts of 'gain' and 'lose'. The dichotomy between development and destruction has proved undebatable beyond doubt.

The herculean feats of economic gains and development have enhanced many times the severity and magnitude of deforestation. The developmental activities of the day in vogue of more modern and advanced exploitive technology without regard to man and his environment are proving ecologically destructive and there is no doubt that such developmental networks have presented various problems of environmental management. It has been felt that modern developmental functioning has dual character of development and destruction. Marriot has rightly compared to-day's 'development' with 'over development' and according to him 'over development' is " . . . the pressing of techniques up to and beyond the point of an optimum relation between man and his environment. In an over

---

* Modified and Revised version of Singh, A.K. and J. Singh's paper (1985) "Deforestation in U.P.: an Environmental threat" in H. H. Singh <u>et al.</u>(ed) <u>Geography and Environment: Issues and Challenges</u>, Concept Publishing Co., New Delhi.

developed area, too many techniques are too exhaustively applied by too many people - to too little land" (Marriot, 1957). And certainly deforestation is one of the outcomes of various developmental activities having the character of extensional development; thus, there is a considerable truth in saying that deforestation is an havoc brought out by thoughtless destructive activities of development (Singh, 1984).

Centuries back when the population was too small, needs were also little, hence the problems of environment were not so acute as to-day. In Uttar Pradesh forest exploitation starts with the Aryan migration which took place sometime around 1500 B.C. "These people were mainly agriculturists. Some of them were also engaged in pastoral pursuits. Forest clearings were made not only for laying out their villages but also for cultivation and pasturage. They cleared forests for the purpose of their settlements. As the population increased more and more, forests were cleared for cultivation, habitation and pasture. They used wooden vehicles such as carts and chariots. They also constructed paths by clearing forests. The villages (grams) of Aryans consisted of thatched cottages. The doors, windows, posts, rafters, ridges and bottoms were entirely made of wood which were obtained from the forests. Felling of forest was resorted to for fuel" (Dwivedi, 1980, p.3). During later Vedic civilization and Epic age, excepting Eastern U.P., there were many large forest tracts spread all over the state. The position of vegetal cover remained more or less the same during the Gupta and Moghal periods also. Gradually with the population expansion, forests began to shrink and this trend continued till the advent of the Britishers, who started clearing forests on commercial basis at a high rate and that commercialized deforestation has continued, leading to a major threat to the environment.

The population explosion (following a 'j' growth curve) and consequent pressure upon natural resources have taken the toll in the northern hilly region of the state - "too much of the actual and proposed development of the Himalayan region is of this character. Himalayas have been regarded as an inexhaustible reservoir of exploitable and expendable resources without regard to the future of the fragile mountain environment..." (Berreman, 1979, p.21). There are many causes and consequences of deforestration. Among various factors leading to deforestation are the activities directly related to development viz., the construction of roads, execution of various river valley projects, extension in agricultural lands, establishment of industrial centres, construction of buildings for various offices and tourist's holiday homes and many more miscellaneous constructions. Table 4.1 depicts the magnitude of deforestation.

Forest fire is also a major problem related to the deforestation. Forest fires have many causes - advertant and inadvertant. Forest fires play a significant role in disturbing the vegetation which further leads to micro-environmental changes, as fires destroy the biotic states of areas under influence. Table 4.2 presents causes and role of forest fires in deforestating the forest areas.

There are several other activities which lead to deforestation. The commercial exploitation of forest resources by the contractors is as important causative factor. "More commonly and rapidly destructive is commercial exploitation by outside contractors who devastate huge areas under government licence for quick profit. They cut forests for lumber, paper-pulp and charcoal. I have seen large forest tracts cleared, the great logs piled high only to be burnt to charcoal" (Berreman, 1979, p.23). The extraction of major and minor forest products by various government and non-government agencies have also adversely affected the forest wealth in various ways. Such commercial exploitations are affecting thoroughly the quantum of afforestation, besides hitting the regenerating capacity of various species at different scales. Among the highly affected species, especially through the extration of timber and Sal, Shisham, Eucalyptus, teak, asna, jamun and khair, etc. (Non-coniferous) and chir, deodar, fir and spruce, kail and cypress (coniferous). Table 4.3 sets out the true picture of species-wise extraction of timber.

From the table 4.3 it is evident that among the coniferous species chir, deodar, fir and spruce are highly affected by commercial exploitation, whereas in non-coniferous species Sal and Eucalyptus are worst exploited.

There are various other usage of forest resources. Among these fuel is most important locally as well as regionally. The various agencies engaged in extraction of fuel from forests are Forest Department itself and purchasers, right holders, concessionists and free grantees. The maximum quantity of fuel wood is being extracted by the purchasers followed by right holders and concessionists. Table 4.4 reveals the structure of fuel wood exploitation by various agencies.

The statistics given in various tables are quite sufficient to tell the saga of speedy exploitation of forest resources. However, such statistics do not explain holistic picture because there are other forms of exploitations also, e.g., lopping,

Table 4.1: **Purpose of deforestation and the area accrued (ha)**

Period	River valley projects	Agricultural purposes	Construction of roads	Establishment of industries	Miscellaneous purposes	Total
1951-52 – 1972-73	93,200	79,700	3,400	19,300	19,900	2,15,500
1973-74	41	178	78	62	13	372
1974-75	02	3,896	333	41	186	4,458
1975-76	312	01	786	-	375	1,474
1976-77	8,732	289	145	23	854	10,043
1977-78	-	-	69	-	8,403	8,472
1978-79	-	07	-	3,999	429	4,435
1979-80	-	-	160	-	123	283

Source : Forest Statistics, Uttar Pradesh, 1983, p.31.

Table 4.2: **Causes of forest fires and area burnt (in ha)**

Causes of fire	Area burnt during			
	1981-82	1980-81	1979-80	1978-79
1. Fire originating in departmental fire conservancy operations	31	4,990	1,003	4,886
2. External fires crossing the fire traces	465	643	562	1,388
3. Fire due to carelessness of outsiders	10,119	21,020	24,271	21,618
4. Fire originating from intention :				
(a) Of turning out games etc.	-	1,739	238	283
(b) Of malice	-	-	134	239
5. Fire due to unknown causes	2,164	36,025	9,455	14,975
Total	12,779	64,417	35,663	43,389

Source: Forest Statistics, Uttar Pradesh, 1983, p.101.

Table 4.3: **Species-wise extraction of timber (cu. m. in round)**

Species	1981-82	1980-81	1979-80	1978-79
A. Non-coniferous	574,034	551,533	690,260	626,014
1. Sal (Shorea robusta)	1,93,669	1,84,257,	2,51,342	2,23,337
2. Teak (Tectona grandis)	9,627	8,110	5,260	3,260
3. Asna (Terminalia alata)	9,187	9,202	13,892	22,265
4. Shisham (Dalbergia sissoo)	15,645	14,924	13,630	14,886
5. Semal (Salmalia malabarica)	2,558	8,313	5,228	4,706
6. Gutel (Trewia nudiflora)	578	476	1,290	2,411
7. Haldu (Adina cordifolia)	4,543	3,145	6,072	9,146
8. Jamun (Syzygium cumini)	5,897	4,862	9,215	11,045
9. Tun (Cedrela toona)	765	1,961	1,625	1,682
10. Khair (Acacia catechu)	3,534	5,349	8,642	24,812
11. Kanju (Holoptelea integrifolia)	2,167	2,842	3,353	5,738
12. Siris (Albizzia procera)	606	262	1,614	276
13. Jhingan (Lannea coromandelica)	1,742	1,801	4,182	2,705
14. Kharpat (Garuga pinnata)	457	530	1,029	756
15. Sandan (Ougeinia oojeinensis)	1,770	1,499	1,535	1,687
16. Eucalyptus	1,14,692	1,88,186	2,06,181	1,61,697
17. Miscellaneous	2,06,597	1,15,814	1,56,170	1,35,605
B. Coniferous	3,73,078	2,44,682	2,71,727	5,23,400
1. Chir (Pinus roxburghii)	3,37,253	2,04,297	2,40,652	4,13,654
2. Deodar (Cedrus devodara)	12,645	7,814	7,052	9,567
3. Fir and Spruce (Abies pindrow and Picea smithiana)	14,155	22,026	13,008	92,869
4. Kail (Pinus wallichiana)	9,089	9,705	9,933	6,849
5. Oak (Quercus incana)	-	772	1,016	369
6. Cypress (Cupressus torulosa)	116	68	66	2
Total	3,47,112	7,76,215	961,987	149,414

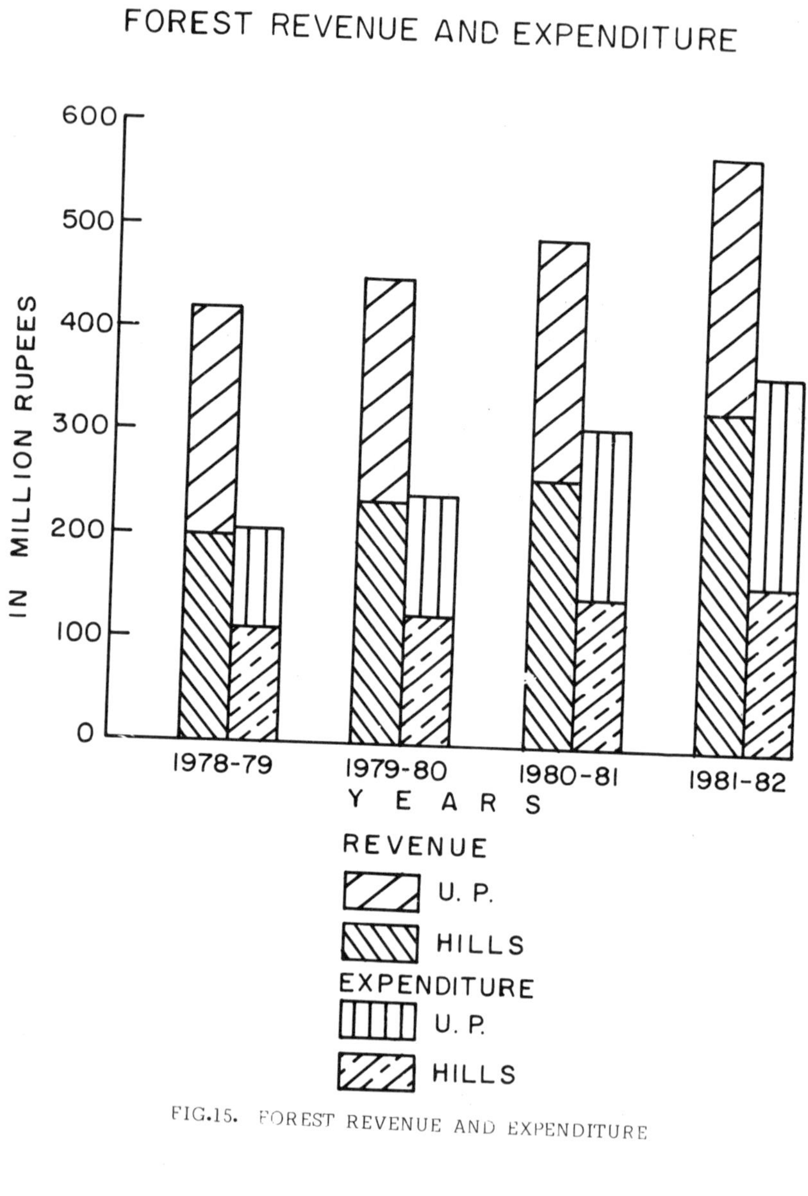

FIG.15. FOREST REVENUE AND EXPENDITURE

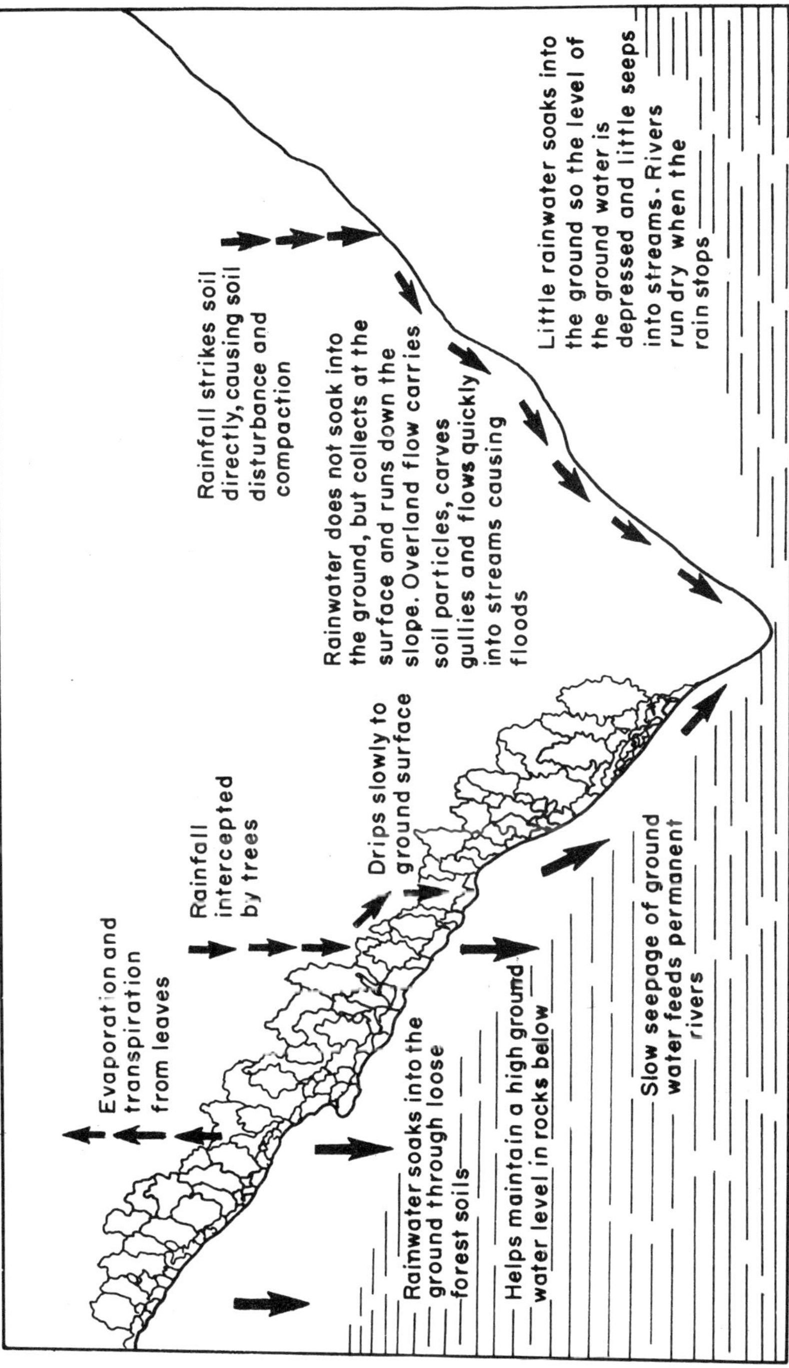

FIG.16. DEFORESTED SLOPES YIELD MORE WATER AND MORE SEDIMENT QUICKER THAN FORESTED SLOPES

Table 4.4: **Extraction of fuel wood by various agencies (cu. m.)**

Agencies	1981-82	1980-81	1979-80	1978-79
Departmental	6,149	582	195	92
Purchaser	10,74,942	10,99,525	16,13,942	15,98,400
Right holders and concessionists	6,61,167	7,48,215	7,11,747	7,17,205
Free grantees	7,548	13,445	33,669	39,577
Total	17,49,806	18,61,767	23,59,103	23,55,274

Source: Forest Statistics, Uttar Pradesh, 1983.

clearing, cutting and extraction of various forest products by local inhabitants illegally and its is quite difficult to quantify these losses. From a recent survey made by us it has been found that at least felling of one tree per annum per family, continues on either for the construction or for fuel. It is also a way of exploitation but is often ignored in government statistics. The average lopping rate by a family per year in the U.P. Himalaya is approximately 15 qts. which is expected to go up in near future. Grazing is also an acute problem and there is a considerable truth in saying that "if we regard pasture and forest as the hill farmer's capital resources which yield a yearly interest in terms of grass, fodder, leaves and humus, then over-grazing and over-cutting of leaves means that the farmers are not only spending their entire annual interest but are also making inroads into their non-replaceable capital resources. This spells disaster, if continued" (Ashish, 1981, p. 27). There is an important aspect of grazing which affects indirectly forests through the reduction in regeneration as "cattle hoofs cut the soil on steep hill sides, leaving it to be washed away in the monsoon rains . . . cause immense damage to tree seedlings and so prevent proper regeneration after felling" (Ashis, 1984, p. 28). Troup (1916) has examined the adverse effects of goats on the survival of chir pine seedlings. He observed that 16 per cent seedlings would be damaged after 8th day by one goat per acre. While 35 per cent would be destroyed after 28th day (Table 4.5). The magnitude of damage becomes high with the increase in number of goats as two goats damage 50 per cent seedlings after 8th day and 88 per cent after 28th day. Likewise, vegetation growth is hampered by other animals too in various ways. Even the position of grazing land in terms of cattle pressure is not satisfactory. Tables 4.6 and 4.7 clarify the situation.

From the Table 4.6 it is clear that the area under grazing amounts to about 80 per cent of the total area under the control of forest department of U.P. of which about 97 per cent area remains open throughout the year for grazing. In terms of animal population, the pressure of grazing is also on the higher side as the whole region records only 1.87 ha. grazing land per cow unit* (1981-82).

Deforestation Causing Threat to Environment

Deforestation leads to not only deterioration in natural environment but also in cultural environment, as all the cultural features and conditions are dependent on the natural surroundings. The natural environment feeds cultural environment and is also in turn fed by the cultural environment. Forests are one of the basic geographical elements and any alteration in forest resources and wealth means alteration in environment. The present study has a limited scope as it is concerned only with certain major problems related with the deforestation.

The forest helps in shaping our natural environment in various ways as it influences micro-climate (by maintaining the temperature, himidity, wind, precipitation radiation), soil environment (affecting soil composition, structure, soil water and chemical properties etc.), flood conditions (through the functionings of evaporation, interception, infiltration, run-off both surface and sub-surface and siltation of river and lake beds, etc.), erosion (by the movement of water or wind) and human health (by protecting physical and chemical contaminations, noise etc.) etc. (Dwivedi, 1980, pp.44-45).

The variation in temperature within and beyond the forest zones is an important aspect ralted with the microclimatic conditions, as "forests exercise a moderating influence on air temperature. The influence is the greatest in a dense forest with heavy foliage" (Dwivedi, 1980, 45). In a simplified way deforestation means increase in temperature. Table 4.8 bears eloquent testimony to the abovementioned fact. If the deforestation continues, we will lose the aesthetic panorama of the area

----

* Equivalence to cow units has been assumed as following (National Commission on Agriculture, 1976), Buffaloes 2, sheep and goats 1/2, Camels 8, other animals 1.

Table 4.5: **The adverse effect of goats on the survival of Chir pine seedlings (After Troup, 1916)**

Number of goats per acre	Year	Percentage of Pine seedlings damaged			
		After 8th day	After 14th day	After 19th day	After 28th day
1.	1912	16	20	27	35
2.	1912	50	68	80	88

Table 4.6: **Forest area under cattle grazing pressure (in ha)**

Area/types	1981-82	1980-81	1979-80	1978-79
Total area under the control of Forest Department	40,68,718	40,67,999	40,58,517	40,80,557
Area open for grazing for the whole year	32,18,502	31,69,142	31,66,981	32,34,825
Area open to grazing for part of the year	85,278	76,677	1,06,470	94,435
Total area open for grazing	33,03,780	32,45,819	32,73,451	33,29,260

Source: Forest Statistics, Uttar Pradesh, 1983.

Table 4.7: **Animal population in Uttar Pradesh**

Animals	1981-82	1980-81	1979-80	1978-79
Cow and bullock	8,86,861	7,71,670	10,10,192	10,46,142
Buffalo	4,20,062	4,48,050	4,32,130	4,87,200
Goat and sheep	3,01,294	3,05,693	2,53,832	3,31,021
Camel and horse	1,26,048	23,744	15,896	9,904
Others	30,802	11,730	21,607	39,763
Total	17,46,037	17,67,761	17,57,514	19,92,783
Cow unit	17,65,067	15,06,887	17,33,657	19,14,030

Source: Forest Statistics, Uttar Pradesh, 1983.

which ultimately will affect the nature and economy of the area. Besides, depletion of different species has different levels of temperature effects. Dabral and his team of workers have concluded that there is a considerable difference in air temperature beneath different species and open places (Dabral *et al.*, 1969) Table 4.8 gives a clear picture of the same.

Table 4.8: **Maximum temperature (°C) of Mussoorie increases with deforestation**

Year	Temperature	Variation (°C)
1950	25.8	-
1955	26.9	+ 1.1
1960	27.2	+ 0.3
1965	28.7	+ 1.5
1970	30.1	+ 1.4
1975	31.9	+ 1.8
1980	32.2	+ 0.3

Source: Negi, S.S. (1982), Environmental Problems in the Himalaya, Dehradun, P. 111.

From the table 4.9 it can be stated that deforestation of chir forests in a local sub-region will lead to an increase at least to 3°C provided other locality factors remain constant. Such impacts of deforestation are clearly visible in various hill areas where the level of deforestation is on higher side. During the summers, due to increase in temperature, use of fans have become usual phenomena in hills. In a recent survey conducted by us, almost all persons belonging to age-group of 70-80 years or more, who have a long expeience of perception of their surrounding, have expressed that usual phenomena of snowfall have become intermittent and summers have beocme painful. Not only by the increasing demands of growing population but by also increase in temperature, the consumption of electricity for demestic needs as well as irrigation is going up, thus affecting other resources of environment.

Table 4.9: **Average air temperature (^{o}C) measured at 14.00 Hours**

Height	Chir plant-ation	Bamboo plantation	Open
Above from 5' ground surface	25.9	26.3	27.3
3' " "	26.0	26.3	27.5
1' " "	26.1	26.3	28.2
6" " "	26.5	26.4	28.5
Ground surface	26.5	26.9	29.6

Wind movement is also related with the nature of species and the density of forest. "Dense forests with heavy foliage reduce the velocity of wind that moves through the forest regions and this reduction of wind movement brought about by forests and shelterbelts adds not only to the comfort of the men and animals but also increases the production of food crops and vegetables growth in leeward side" (Dwivedi, 1980, p. 46). There are considerable experiences of such effects in the hill districts of the state under study.

Deforestation also affects the soil and air moistures, as these are related with the temperature and wind movements. The precipitation is also a related phenomenon, but a little bit controversial. The common belief is that the forests augment the intensity of rainfall. In this connection Blandfords example may help to realise the impact of forest on rains (as mentioned by Dwivedi, 1980, p. 47), Blandfords has compared the data of fourteen stations of central province of India during the period 1867-75 with the data of the 1876-85 period. He observed that mean annual rainfall increased in the latter period by 173 mm or about 12 per cent because of afforestation over there.

Forests also help in the formation and composition of soils by adding humus and influencing other soil-forming factors such as temperature, rainfall etc. Soil erosion is also directly related with the forest coverage to a certain extent. From the surface which has a forest coverage, the amount of run-off will be less than that from open areas having no foest coverage (fig.16). The flow of water (in terms of amount and velocity) influences the amount of soil erosion. Forest role in moderating the floods is beyond doubt. From the table 4.10 it has been observed that the severity of floods in terms of frequency and loss of lives and property is increasing (Table 4.10). The damage caused by floods during recent years (1976-78, i.e., 2 years) is more than two times from that of earlier twelve year span (1953-65). This is not only even the share of U.P. in flood damages is much high than their counterpart flood affected states (Table 4.11). Report of the National Flood Commission (1980) reveals that during 1950-54 the loss of houses and cattle were 113,000 and 4,541 respectively whereas the total loss of human lives was 176 (Table 4.11).

The National Flood Commission is of the opinion that among the various causative factors leading to floods in the Gangetic plains, deforestation is supreme (National Commission on Floods, 1980). The major river of the state - the Ganga originates from the hilly tracts of Uttar Pradesh. This river system covers approximately two-third area of the state which often experience floods and at least one major flood in each and every decade. Keeping in view the major flood disasters of the present century, Ehrich has remarked, "......was the flood disaster of 1978 not enough to teach this country a lesson, or will the problem increase even more and those, who have not taken the needed decision in time, would be in public disgrace? Should we say that this country needs even much bigger floods to awaken those who decide national budget priority. If this be so, let more time pass by unheeded, the next flood would be knocking at the door" (Ehrich, 1980).

The role of forests in retaining water is also important especially in the hill areas as the springs are fed by such retained water (Singh and Rawat, 1985). Deforestation has already affected spring environment in several ways. The concentration of springs in hill areas of the state is directly, related with the rock structure of the region and the forest species lying over there.

Table 4.10: **Major trends in area affected and damage by floods (area in lakh hectares and damage in lakh rupees at 1952-53 price)**

	INDIA				UTTAR PRADESH			
	1953-65	1966-70	1971-75	1976-78	1953-65	1966-70	1971-75	1976-78
Total area affected	59.62	75.10	91.08	175.70	17.44	20.12	23.46	40.90
Cropped area affected	20.99	33.58	46.86	86.60	7.93	10.42	13.22	22.33
Non-cropped area affected	38.63	41.52	44.22	89.10	9.51	9.70	10.24	18.57
Total damages	5,123.45	9,223.74	15,238.25	25,954.54	1,270.58	1,637.22	2,728.05	7,338.65
Crop damages	3,880.41	6,148.17	10,160.15	15,678.50	1,055.56	1,296.27	1,748.37	5,249.20
Non crop damages	1,243.04	3,075.57	5,078.10	10,276.04	215.02	340.95	979.68	2,139.45

Source: National Commission on floods, Ministry of Energy and Irrigation, New Delhi, 1980.

Table 4.11: **Share of major states in Flood damages (at current prices) in recent years (in percentage)**

States	1971	1972	1973	1974	1975	1976	1977	1978	Adverage for 1971-1978
Andhra Pradesh	-	12.0	(a)	(a)	(a)	23.9	50.8	(a)	15.4
Assam	(a)	15.3	2.9	3.4	(a)	1.4	2.60	(a)	2.1
Bihar	34.4	1.1	3.5	67.1	56.5	23.2	1.0	16.7	23.9
Gujarat	(a)	(a)	20.9	(a)	15.3	6.6	4.5	1.2	5.7
Haryana	1.0	(a)	(a)	-	(a)	4.3	5.9	5.3	3.2
Madhya Pradesh	(a)	(a)	12.2	(a)	(a)	(a)	-	-	1.3
Orissa	5.0	29.6	9.0	(a)	10.0	(a)	(a)	3.3	4.5
Punjab	(a)	(a)	1.4	(a)	2.1	5.6	(a)	(a)	1.4
Rajasthan	(a)	6.9	17.8	(a)	4.5	2.3	6.3	1.2	4.5
Tamil Nadu	-	17.3	4.7	-	-	(a)	12.7	-	3.8
Uttar Pradesh	27.6	2.4	14.4	18.5	8.3	27.8	8.1	56.0	23.8
West Bengal	25.2	9.5	7.8	4.7	(a)	2.4	3.3	7.1	7.0

(a) = Less than 1.0 per cent

Source: National Commission on Floods, Ministry of Energy and Irrigation, 1980.

Table 4.12: **Damage to crops and houses due to floods in U.P.**

Year	Area under floods (Lac of crores)	Damage of crops (Rs. crores)	House damages (Rs. thousands)	Damage of houses (Rs. crores)	Cattle lost	Human life lost	Total loss (In cr Rs.)
1950	26.00	12	25	1.25	2,200	128	13.25
1951	No floods	-	-	-	-	-	-
1952	No appreciable floods	-	-	-	-	-	-
1953	22.51	10	80	4.0*	341	23	14.00
1954	17.00	7	8	0.4	2,000	25	7.40
Total five years	65.51	29	113	5.65	4,541	176	34.65
Average	13.00	5.8	22.6	1.13	980	35	6.93

* Assessed at Rs. 500 as suggested by revenue sect. U.P., on phase

Source: National Flood Commission, Vol.II, 1980, p.147.

Deforestation Vs. Environmental Degradations : Experiences and Examples

It is worth quoting "once beautiful, forested hills are becoming bare, ugly landscape . . . The Himalayas says Eric Eckholm, form an ecological Gibralter whose fate looms over the well-ebing of hundreds of millions - about 50 million in the mountains, and another about 300 million in the Indus, Ganges and Brahmputra valley systems. The ecological clock is tickling away. The deserts of Afghanistan are in the late evening; Bhutan and Arunachal Pradesh may still be in the early morning; and the Central Himalaya is in the late afternoon of the Himalayan days. Wildlife is being fast exterminated, and both man and mountain are becoming poorer". This statement expresses the experiences and the happenings in the hills. Further, he has also expressed his fear, "Man has been the maker of deserts in past ages, even with bare hands, and often in the name of progress. Now that he is multiplying fast, and has more powerful technoloy, his capacity to endanger the balance between Man and Nature ('only one Earth') has been enhanced many times" (Moddie, 1981). Berreman, an anthropologist, has given a very sincere remark regarding the development of Himalaya putting it 'more inclined towards destruction rather than development'. The green revolution has various positive role in improving the well-being of people, at the same time its negative impacts from the environmental stand point may not be separated from the degradational studies of the environment in general and on forests in particular. There is no doubt in saying that the 'Green Revoluation', has become the eater of vegetal cover. In consonance, the replacement of 'nature's greenery by crops' has dual character of both the development and destruction. Hence, intensification is rather more imperative than extensification, keeping the environmental loss of vegetal cover. Side by side the replacement of 'natural greenery by artificial colouring' should be checked (Singh and Singh, 1985). For the pursuance of short-sighted well-being people have jeopardised their future. "The deforestation and expanding unscientific (not based on land capability) agriculture in the hills are causing massive environmental degradation and untold misery" (Sing and Chaturvedi, 1981). In the various localities of the Himalayan region "orchides are getting scarce and rare due to indiscriminate felling of forests" (Maheshwari, 1980). Disturbances in wild life status, alteration in herbal vegetation and microbial community, increasing noise and mechanical disturbance to nestling birds are major issues related with forest environment. "In the past few decades, the growig impact of human population, in the form of deforestation, coversion of forest land into agricultural fields, invasion over steep slopes .

. . has brought irreversible ecological changes . . . the soils of this zone are being exported to the down hill regions without the latter paying compensation" (Bisht et al., 1984). The statements mentioned above, are enough to portrary the human feeling of 'inhuman activities of human beings' in the name of development.

Now we may discuss,. how the deforestation is hitting human life. On the day, when the nation was celebrating its 30th Independence day (15th August, 1977) the inhabitants of Khela village-cluster situated on the Indo-Nepal-Tibet border, were crying due to land-slide caused by deforestation and about forty-four persons had died along with the loss of hundreds of cattle. Alakhnanda episode (Table 4.13), Tawaghat tragedy and various other calamities have eaten up not only the wealth of thousands inhabitants but also their lives.

Table 4.13: **Loss of lives and property accrued to floods and landslides in Alaknanda valley**

Year	Loss of lives		Loss of property (in lakh Rs.)
	Human	Cattle	
1971	10	422	10.77
1972	7	22	1.25
1973	3	169	8.68
1974	9	25	0.77
1975	1	85	7.05
1976	1	21	7.00
1977	3	22	7.75
1978	14	164	138.96
1979	45	143	36.14
Total	93	1,073	218.37

Source: Pathak, C.C. Pahar, Nainital, 1983, p.126.

Among the various examples of slope failures in the hill regions of Uttar Pradesh, Raipur, Kalagarh and Malota Nallas, Kherwa, Garampani, Kaliasaur, Nandprayag, Belakuchi and Kermi slope-failures are important. The accumulation of materials caused by landslides and debris has been observed by various authors (Nossin, 1967: Negi, 1982; Haigh, 1984) which pose alarming situation. The activities of road construction in the Himalayan region is perhaps one of the major causes of landslides and debris slides. According to Valdia's estimate, the 44,000 km long road constructed in the hills has given rise to

2,650 x 10^6 m^3 of debris. Each year landslides and rockfalls generate 550 km^3 of debris per km of the road, causing 24 x 10^6 m^3 of sediment to slide down the slopes killing vegetation and chocking mountain streams (Valdia, 1985). Thus, introduction of new roads in the area may lead to problems of high dimension.

The danger of landslide in Nainital town is a growing concern. On the slope surfaces where the severity of landslides is often expected, the constructional works of living apartments are inviting such disasters in a big way. It is an alarming situation. Centainly such sides of slopes must be avoided for living purposes (see also Mishra and Pant, 1982).

An intersting account of the area affected by various kinds of slope failures has been presented by Singh (1984). Table 4.13 shows that in all the cases area from where slope failures occur is less than the receiving end. Intensity of vegetational loss depends upon the volume and nature of out falling material, angle of slope and height of places from where slope failures take place.

Soil erosion is a severe problem throughout the state and deforestation is one of its causes. In the Himalayas, the problem is more severe on account of massive deforestation. In several localities of the Kumaun and Garhwal Himalaya, the incidence of soil erosion is very frequent and common. Owin to the high degree of slopes and sufficient rainfall, whole of the U.P. Hill region provides an increased opportunity for fluvial actions which ultimately pave way to erosion. There are various spectacular examples of this process which are neutralizing almost all the conditions related with soil formation and associated activities. According to one study, the Siwalik ranges are comosed of Mio-pliocene sands and gravels which seem to be erosion products from the Himalaya (Haigh, 1984, p.195). Dani *et al.* (1982, pp.184-188) have estimated the proportion of different erosional classes in some of the major landuse categories of the five sub-catchments of Ramganga catchment (Table 4.15). From the table it has been inferred that the rate of high erosion has positive correlation mainly with the slope and improper management of land. In another study which has been conducted in four micro catchments, having 28^0 to 40^0 slope, in the Garhwal Himalaya reveals - "alluvial fan deposits collecting beneath two deforested catchments were 5-7 times larger than those collecting beneath two comparable forested catchments and they contain a relatively high proportion of sandsized particles" (Haigh, 1982).

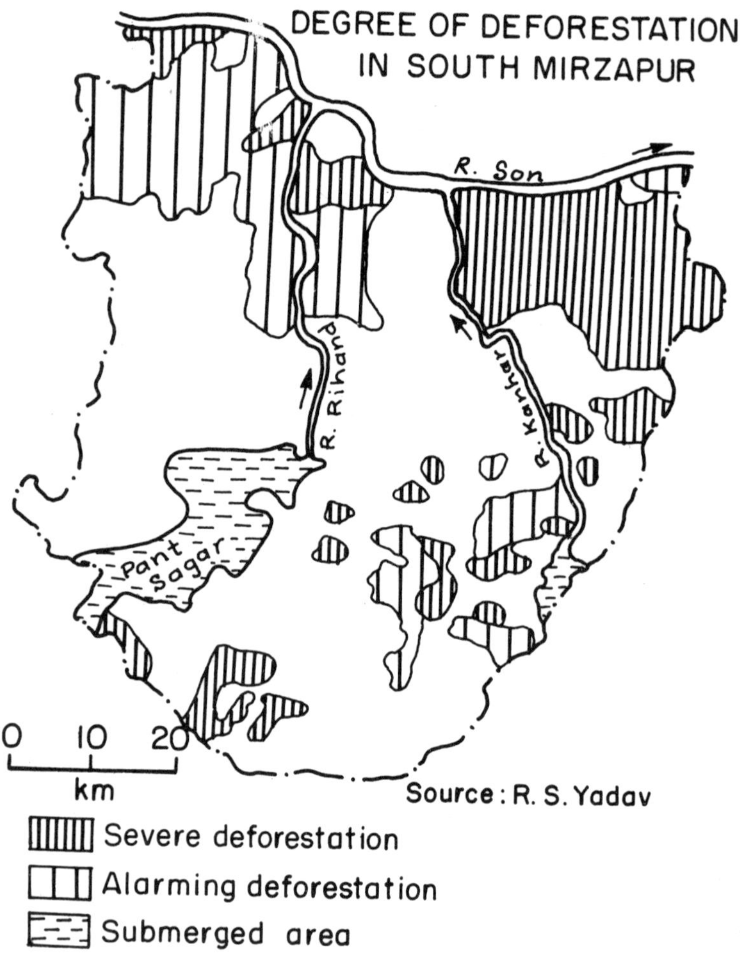

FIG.17. DEGREE OF DEFORESTATION IN SOUTH MIRZAPUR

Table 4.14: **Slope failure and spread of materials**

Localities	Number of observations	Actual* areal coverage of slope failures (sq. m.)	Spread of Materials (sq. m.)			
			Total area	Highly affected zone	Moderately affected	Less affected zone
1. Along Kathgodain Nainital Road	5	127	199	71	87	41
2. Along Bhawali Almora Road	7	192	258	108	99	51

* Average, depth 10 cm.

Source: Singh, A.K. "Slope failures vs. Material spread in Himalayan Environment", 1984 Unpublished.

Table 4.15: **Land classification with their distribution in various Erosion Intensity Classes**(in percentage)**

Subcatchment	Major land	Erosion intensity classes* E_1	E_2	E_3	E_4	Per cent[2]
Nair	A	42.28	52.84	0.81	0.07	37.26
	B	17.60	76.15	5.50	0.75	38.24
	C	73.71	26.29	-	-	10.93
	D	4.91	75.58	19.51	-	0.08
	E	-	-	-	-	4.54
	F	-	-	-	-	8.95
Middle Ram Ganga	A	59.12	40.77	0.11	-	30.92
	B	38.32	54.48	5.87	1.33	37.55
	C	68.02	28.80	5.18	-	10.10
	D	100.00	-	-	-	0.29
	E	-	-	-	-	5.24
	F	-	-	-	-	15.90
Badangarh	A	45.26	53.15	1.15	0.44	32.60
	B	28.17	46.04	21.94	3.85	40.24
	C	92.04	6.99	0.13	0.64	6.34
	D	60.74	-	-	39.26	0.03
	E	-	-	-	-	4.42
	F	-	-	-	-	16.17
Haldgarh	A	31.67	60.50	7.83	-	17.10
	B	68.28	14.47	21.63	0.62	23.42
	C	86.55	10.48	2.97	-	6.50
	D	-	71.62	28.38	-	0.02
	E	-	-	-	-	3.06
	F	-	-	-	-	49.90
Mandal	A	31.86	59.99	7.64	0.51	21.20
	B	40.36	20.12	35.23	0.29	29.60
	C	65.00	20.09	14.76	0.15	9.13
	D	21.74	47.18	25.18	5.90	6.71
	E	-	-	-	-	2.77
	F	-	-	-	-	36.59

1. Percentage of Erosion intensity are given to the total area of each land use category in each sub catchment separately.
2. Percentage of each landuse category to the total geographical area of each catchment area.

** Basic data obtained from soil and conservation Division of Forest Department

A = Cultivated land; B = Cultivated waste and community grazing lands;

C = Community forest; D = Steep zone pasture land;

E = Miscellaneous area; F = Reserve Forest.

E_1 refer to very slight erosion conditions in the form of splash to sheet erosion, E_2 refers to a relatively higher degree of deterioration in the form of sheet to rill erosion in a more sloping land while E_3 and E_4 denote sill higher degree of erosion in the form of rills and big gullies.

Extensive degradation of land has occurred in the forms of ravines along the banks of the Yamuna and Chambal rivers. In U.P. alone about one-third (12.30 lakh ha) of the total ravine affected area lies. The probable causes of ravine development are loose and friable nature of the soil, steep slopes, undulating terrain, faulty agricultural practices, illicit cutting of trees and bushes etc., (National Commission on Agriculture, 1976, p. 129).

Due to deforestation, the badly damaged compartments of Syamkhet, Karil, Ruinsi and Mahrora of Nainital and Amri, Puraban Palbelon of Pithoragarh are experiencing more erosion (10-15% higher) than previous levels. The same situation prevails in the various forest areas of Almora, Garhwal, Mussoorie, Chakarata and Tehri-Garhwal districts where deforestation is comparatively on higher side. Inhabited zones lying along the highways are experiencing a great havoc from soil erosion. In the southern part of the state, the situation of soil erosion is also alarming. The map regarding ecological damages in South Mirzapur clearly depicts that more than 60 per cent area of Son-Kanhar interfluve is severely damaged while the tragedy of deforestation is alrming in the north west part of the Son-Rihand interfluve along with various patches stretched over south-eastern part of the area (Fig.17). "The ruthless cutting of forests and its conversion into agricultural land has led to gradual increase in volume of run-off . . . led to increase in the extent of soil erosion" (Yadav, 1984).

In almost all the districts of Uttar Pradesh (leaving Hill districts) the deforestation has invited a new problem on soil environment. Previously when the woodlands and forest pathces were available in abundance throughout the region, the cattle dungs were utilised as organic manure because the fuel needs were ful-filled by the jungles available nearby each and every vilage. But nowadays, the situation is entirely different. Such woodlands and forest patches have disappeared due to speedy extension of agriculture to meet the increasing demands of rapidly growing population. Thus, to meet the fuel requirements, cattle dungs are being used frequently in the form of dried cakes. From the southern districts of the Plateau region to northern districts of Tarai region, soil is suffering a great loss of organic manure. From a recent observation, it has been found that in few districts of Eastern Uttar Pradesh, about 80 to 90 per cent of total cattle dung produced are diverted to kitchen fuel rather than field manure. The situation varies as per the availability of wood land. The condition in southern, west and tarai region of U.P. is comparatively better. Social forestry schemes can help much in solving this organic loss accrued to the soil. On the one hand we are providing little organic manures to our fields, on the other the population pressure on the land has increased. Due to the agricultural extension programme and intensive farming, the crop cultivation intensity has gone up and at the same time the use of cattle dung manures has come down to a great extent.

Due to the high rate of siltation (much higher than expected), the age of Ramganga Dam Project has been calculated at 76 years against the previous estimated age of 100 years, certainly due to the misery of erosion (mithal et al., 1984) as a consequence of deforestation.

Soil erosion, by-product of deforestation, retrogrades the wheel of agricultural economy by reducing the fertility of soil. When the soil moves down, the economy based on soil also follows it. Thus, the role of forest wealth in maintaining 'soil environment' is unquestionable, and if deforestation occurs (in the study area deforestation is an acute problem) or continues the recurring loss in soil system is inevitable.

In hill areas of Uttar Pradesh, forests are performing a big role in maintaining the 'spring environment'. The water retaining quality of forests and further its gradual discharge directly governs the nature of springs in terms of quality, quantity and periodicity of springs (Singh and Rawat, 1985). The periodicity as well as the number of springs are also affected by different species in Kumaun and Garhwal region (Table 4.16 and 4.17).

Table 4.16: **Concentration of springs**

Region	No. of springs observed	Forest zones with oak	Forest zones without oak
Kumaun	60	43	17
Garhwal	60	47	13
Total	120	90	30

From Table 4.16 it is evinced that the concentration of springs is more common in and around oak forest zones than non-oak forest zones. Besides, the gradual depletion of oak forests has also adversely thwarted the periodicity of springs. In many areas where the depletion of such forests is high, the springs generally dry-up while in some cases (if not dried up) a significant decrease in water discharge is found. Valdia has also pointed out the drying of springs due to deforestation. He has examined springs of Gaula River Catchment in Nainital district and found 45-46 per cent springs have gone dry or become seasonal (Valdia, 1985, p.29). Because of this, nowadays, hill areas face acute shortage of water during summers. The hillmen could not manage adequate quantity of drinking water in this season. Thus, it can be stated that springs are responsible for shaping/dishaping the economy and habitat of this region.

Besides this, the replacement of oak forest with pine in the Himalaya is also affecting the nitrogen cycle (Table 4.17) as 'the pine is a light-demanding, fire adapted but fire promoting species. The surface fires averaging once every 2 or 3 years cause substantial nitrogen losses in the pine forest (Singh, J.S. et. al., 1984).

Table 4.17: **Nature of springs**

Region	No. of springs observed	Extinct	Seasonal	Reduction in discharge
Kumaun	60	10	18	32
Garhwal	60	12	22	26
Total	120	22	40	58

Source: Singh and Rawat, 1985, p.46

Retreat of Glaciers in the Himalayan region is also an important aspect from the environmental point of view and is related with the climatic conditions of the areas. Though the impact of deforestation on glaciers has not been studied scientifically, yet it has been argued that deforestation as the agent of change in micro-climate may be one of the factors leading to the retreat of glaciers (Table 4.18).

Table 4.18: **Retreat of glaciers**

Glaciers	Period	Year	Retreat in mts.
Milam	1849-57	108	1,350
Pindari	1845-66	121	2,840

Source: Vohra, 1981, p.141

Areas endowed with rich natural vegetation attract wild animals and any loss in vegetal cover compels them to migrate due to the force of ecological fugacity. With the depletion of forests in the state, many species of wild animals and birds have become rare or are extinct. On the one hand, deforestation as a repulsive factor (by reducing the opportunities of security) disturbs the habitation of animals whereas on the other, it also provides opportunities to the hunter by providing better conditions of inside mobility. In different locality of Kumaun region there are various species which are under the state of trouble (Table 4.19) in various proportion. This all has happened due to onslaught on forests by different agencies. The second important problem is related with the expansion of various Xerophytic species in the Kumaun Himalaya. Recently authors have examined a rapid growing species of Euphorbia over an area of 10 km^2 in Kabri Ghat (Nainital), while enroute Kathgodam and Bhowali, the fast extending species of Agave also attracts the attention of environmentalists. The expansion of these Xerophytic species will also lead to a change in the character of associated flora of these areas not only in the concerned areas but also in the surrounding areas.

The preceding paragarphs bear eloquent testimony to the fact that deforestation has posed a great menace to environment of the state.

Table 4.19: **Dry matter and nitrogen (N) abundance for the tree layer of pine and oak forests***

	Pine forest		Oak forest	
	Dry matter	N	Dry matter	N
Compartment Pools (tonne ha^{-1})				
Vegetation	210.50	1.106	377.10	3.299
Litter	10.62	0.131	5.70	0.094
Soil (to 20 cm)	-	3.964	-	5.819
Total	221.12	5.201	382.80	9.212
Accumulation in net production (tonne ha^{-1} yr^{-1})	16.70	0.143	15.10	0.185
Reabsorption before senescence (tonne ha^{-1} yr^{-1})				
Foliage	-	0.045	-	0.012
Wood	-	0.008	-	0.006
Total	-	0.053	-	0.018
Net uptake from soil (tonne ha^{-1} yr^{-1})	-	0.091	-	0.167
Return (tonne ha^{-1} yr^{-1})	7.65	0.060	7.08	0.107
Retention in vegetation compartments (tonne ha^{-1} yr^{-1})	9.05	0.031	8.02	0.060

* Received with courtsy from Prof. J.S. Singh, Department of Botany, B.H.U.

Table 4.20: **Species under the state of trouble**

Name of the species	Locality	Altitude (feet)	Levels* A	B	C	D
Psilotum nudum(Linn.) Beaux	Almora (along Kosi River)	4,000				✓
Podophyllum hexandrun Royle.	Pindari	12,000				✓
Thylacospermum rupi-fragrum	Sunderdhanga Glacier	11,500				✓
Balanophora involucrata R. Br.	Above Dwali (enroute Pindari)	11,000		✓		
Paeonia emodi Wall ex. Royle	Doonagiriekhati	Above 7,000	✓			
Heteropogon malanocarpus Benth	Almora	5,000	✓			
Ephedra gerardiana Wall	Pindari	12,000		✓		
Herminium mackinnonii Duthie	Binsar	6,000		✓		
Orobanche aegyplics Pers	Syldey	-	✓			
Nardostachys jatamansi	Pindari	10,000-12,000			✓	
Picrorhiza kurrooa Royle ex Benth	Pindari	-do-			✓	
Aconitum heterophyllum ex Royle	Pindari	-do-			✓	
Blyya auberti Rich	Garur	2,500		✓		
Hydropryum griffithii (Wall ex Graffith Taulsne)	Hawalbagh	1,200		✓		
Hemiphilia cordifolia (Linndle)	Shikhar	7,000	✓			
Botrychium virginianum Sw.	Jageshwar	6,000		✓		
Osmunda regalis Linn.	Khati	8,000-9,000		✓		
Datisca cannabins	Almora	4,500		✓		
Peranema cytheoides	Kapkot	6,000		✓		

A = Endangered; B = Rare; C = Vulnerable; D = Threatened

1. Received with courtsy from Dr. P.C. Pandy, Department of Botany, Kumaun University Campus, Almora.

FIVE

Managing Forest Resources

The continuous onslaught on forest resources and the consequent massive environmental degradation in terms of floods, soil loss, landslides, lives and property etc., on the one hand, and the desirability of sustained yield of goods and services on the other, call for steps towards rational management of forest resources. However, it is very simple to talk about the management of forest resources but the reality is somewhat different. From the point of view of layman, forest resource management is to stop or check the deforestation and to promote afforestation. Though these two related issues act as stepping stones in the management of the forests, there are several other issues, which should be taken into account. A parallelism in afforestation and deforestation has put the state on the same footing as it was 17.11 per cent forest area in 1971-72 and 17.41 per cent in 1981-82. In fact the real situation is more severe as the rate of deforestation is on the higher side as compared to afforestation.

Before proceeding towards the tenets of management of forest resources, the clarity regarding what is forest management is necessary. If people are unaware of the desirability and aims of forest management, there is little or even no hope to achieve the targets of management. Therefore, let us begin with 'what' rather than 'how' to manage the forest resources. In a very simplified version the management of forest resources is basically 'the management of forest eco-system'. The forest eco-system encompasses all micro and macro features, conditions and influences of the environment that affect growth and perpetuity of forest. Dwivedi (1980, p.353) has rightly emphasized that the forest resource management involves the management of (i) forest land, (ii) forest flora, and (iii) forest fauna. The authors feel that few more variables like management of population both of cattle and of human, need-system, awareness, carelessness of resource managers as well as of inhabitants, politico-legal complexities, etc. should be considered.

From the geographical stand point we broadly categorise the

forest management imperatives into:

(a) Management of naturual environment, and

(b) Management of cultural environment affecting and being affected, directly and indirectly by the forest eco-system. The following variables are taken into account in the management of both the physical and the cultural environment related with forest resources.

MANAGEMENT OF NATIONAL ENVIRONMENT

Natural environment consists of natural features, influences and conditions. The life-system is the replica of prevailing conditions of the natural environment upto a certain extent. Natural environment is composed of climatic conditions, soil and water status, vegetational conditions, and land characteristics etc. All these singularly as well as collectively affect the nature, growth, propserity and so many things related with forest eco-system.

(1) Management of soil environment

The role of soil in determining the growth of natural vegetation hardly needs emphasis. The management of soil environment is an urgent need of the day bth from the qualitative as well as quantitative perspectives of forest resources. Soil can be managed in many ways through different techniques and also by various approaches. The first and foremost aim of soil management is to check the heavy loss of soil. In the area under study, the problem of soil erosion has become very critical specially in the hill areas of Uttarakhand and in the southern plateau region. The problem of soil erosion is also rather acute in the Gagetic plain (Chapter IV). Control over rapid soil runoff can be exercised through the following measures:

(i) Terracing: The most common practice to control the soil erosion in the hill areas is terracing of lands. The practice of terracing requires some basic knowledge of surface gradient, nature of parent rock and the intensity of rainfall. Despite low density of population, the regions of steep slope with high inten sity of rainfall face acute problem of soil erosion, hence it is advisable to make inward terracing in such areas instead of outward and flat terracing. It has been observed that people of hill areas, in a few cases, intermingle their plots to have a bigger size located at terraces of marginal height (0.3 - 0.5m) but in long-term it paves the way for soil loss. Thus, this short-sighted approach should be avoided.

(ii) Gully checking: Gully actions cause little erosion in initial stage, but after sometime with increasing gully action soil erosion increases tremendously. Thus, it is a must to prevent the process of gully formation at the initial stage itself. However, the main technique to control gully action is plugging through boulders, pebbles and other resistant earthen materials but it is feasible only for smaller areas and smaller gullies. So far the broader areas/gullies are concerned, it is a costly and impracti- cable device. Planting fast growing native herbs and shrubs may be taken as an alternative to check the expansion of gullies. But care should be taken to plant seedlings from the origin of gullies not from the midway, otherwise it may be a futile exercise only.

(iii) Check dams: The quantum of soil erosion can be reduced to a considerable extent by constructing check dams at suitable sites and elevations. The number and size of check dams should be based on the water catchment area. Besides, they should be constructed in a series at successive height intervals rather than in isolation.

There are a few more techniques of soil conservation among which training of streams into narrow channels and afforestation are important ones. The afforestation has been dealt with in detal, later.

(iv) Nutritional management of soil: Soil is the basic source of nutritional stock and the amount of various organic and inorganic constituents present in the soil directly affect the amount and growth rate of different and divergent forest species. Thus, by doing the nutritional management of the soil by alteration or additions of proper organic/inorganic nutrients, the targets of forest resource management can be achieved to a certain extent. For the purpose of better output of the forest flora, manuring and chemical fertilization are necessary. Persons engaged in such activities of soil management must be aware of the situation regarding the existing budget of soil nutritional state. To improve the fertility of the soil, the basic need is to prepare a macro, meso and micro level chart of soil nutritional surplus and deficit conditions along with the listing of forest species growing therein and their nutritional requirements. On the basis of such charts/maps it can be easily decided where, how much and what kind of nutritional amendments are necessary. Thus, for this purpose not only the micro level soil testing is necessary, even the knowledge of optimal nutrient requirements of different species is needed. There are two basic groups of materials to be assessed for soil nutrition management : organic matter and inorganic chemical fertilizers.

Among the major organic manures are the various combinations of compost, oilseed by-products, waste materials, decomposed fish, green crops, cattle excreta etc. Table 5.1 illustrates the average nutritional value of different organic manures.

Table 5.1: **Average manurial value of different organic manures**

Sl. No.	Name of the manure	Nitrogen	Phosphorus acid (P_2O_5)	Potash K_2O	Calcium CaO	C/N ratio
		Percentage of nutrients				
1.	Farmyard manure	0.50	0.30	0.50	0.30	20
2.	Compost	0.75	0.50	1.50	2.50	15
3.	Night soil	1.50	1.00	0.50	0.30	7.5
4.	Town refuse	0.50	0.30	1.00	0.40	20
5.	Animal blood	2.60	0.20	0.10	0.40	3
6.	Slaughter house waste	1.50	0.55	0.40	0.46	2
Oil Cakes :						
7.	Castor cake	4.35	1.85	1.40	-	10
8.	Cotton seed cake (undecorticated)	4.00	1.90	1.62	-	7-12
9.	Cotton seed (decorticated)	6.40	2.90	1.15	-	-do-
10.	Ground nut cake	7.30	1.50	1.35	-	-do-
11.	Karanj cake	3.95	0.95	1.32	-	-do-
12.	Linseed cake	5.55	1.45	1.28	-	-do-
13.	Mahua cake	2.50	0.80	1.85	-	-do-
14.	Neem cake	5.20	1.08	1.45	-	-do-
15.	Sun flower cake (Undecorticated)	4.90	1.45	1.23	-	-do-
16.	Sun flower cake (decorticated)	7.85	2.20	1.92	-	-do-
17.	Sesame cake	6.20	2.10	1.25	-	-do-
Other Organic Manure :						
18.	Bonemeal (Raw)	3.00	20.00	-	30	-
19.	Bone meal (steamed)	7.00	22.00	-	32	-
20.	Fish manure	7.50	5.55	1.0	-	4
21.	Guana (dried excreta of sea birds)	12.50	10.00	2.00	4.00	-
22.	Press mud (by product of sugar industry)	1.50	4.55	4.5	4.50	-

Source: Dwivedi, A.P. Forestry in India, Dehradun, 1980, p.280.

Among the chemical nutrients, nitrogen, phosphorus, potassium and calcium are the most important for the replenishment of soil deficiency. Application of these elements facilitates a fast recovery of soil nutritional levels. Dwivedi (1980, p. 283) has presented information on suitable chemical fertilizers for different types of soil (Table 5.2).

Table 5.2: **Suitable fertilizers for different types of soils**

Sl. No.	Name of the fertilizer	Soils for which suitable
1.	Sodium nitrate	Good for acidic soils, it should not be applied in soils with poor drainage where Na^+ ions tend to accumulate. It may be used in light soils. It is easily leached.
2.	Calcium nitrate	It is suitable for acidic soils. Nitrate nitrogen is easily leached down.
3.	Anhydrous ammonia	Very good for acidic soils. Loss of nitrogen is more in light soils which have low clay and organic matter content. The loss may go upto 33% under certain conditions. Application of small quantities of organic matter in the soil helps the retention of ammonia in light soils.
4.	Ammonium sulphate	It is good for neutral, calcarious, saline and alkaline soils. It can be very effective in acidic soils also but for short term use. Its long use in acidic soils may deteriorate such soils and availability of calcium, phosphorus and potassium is reduced. It can be used effectively in light and heavy soils and in wet and dry soils.
5.	Ammonium sulphate nitrate	This is an acid forming fertilizer and its suitability to soil types is almost similar to ammonium sulphate. For heavier soils this is comparatively better.
6.	Ammonium nitrate	It is suitable for alkaline and neutral soils. It is the best fertilizer for dry areas.

7.	Calcium ammonium nitrate	This can be applied in the soils of different reaction. It should not be applied in light soils where leaching is heavy.
8.	Ammonium chloride	This is an acid forming fertilizer. It should not be applied in soils which are rich in chloride salts, otherwise its application may cause chloride toxicity.
9.	Mono-and diammonium phosphate	These should be used in well aerated soils. These are good for leguminous species. These can be safely applied in acidic, alkaline and neutral soils.
10.	Urea	Urea should be applied in well-aerated dry soils. It should be carefully applied in sandy soils because it is likely to be leached down.

Source: Dwivedi, A.P.Forestry in India, Dehradun, 1980, p.283-84

(v) Soil moisture and water management: Soil water is the basic ingredient for seed germiation and for the sustenance of plant growth. On account of a canopy of tree leaves and an under growth of grass, seedling and herbs along with a dense litter of leaves and twigs, forest floor, water loss through evaporation is reduced. Further, permeability is increased, a significant amount of water infiltrates into the soil and surface run-off of water is reduced. Excess water in the sub-soil is transpired and waterlogging is prevented. Thus, through proper afforestation programmes, soil water of a given area can be maintained and even enhanced and higher levels of natural regeneration target can be achieved. Through the process of organic manuring also, soil moisture relations can be improved.

(vi) Management of soil formation process: Soil formation is in herently a very slow process. On account of active and increased erosional action, accelerated downward movement of soil has jeopardised the formation and sustenance of soil mantle. This has become a very common feature all over the Himalaya. Through various measures of soil protection, conditions for soil formation can be promoted to a considerable extent. Protection redevelops the plant cover and the associated soil micro-organis mic complex, aiding in accumulation and aggregation of soil particles into an integrated body.

(2) Management of slope failures

Slope failures are the basic problems of the Himalayan region. Among the main causative factors—geological conditions, slope and gradient, hydrological conditions and biotic activities are worth mentioning. Emphasis should be given on the role of vegetative component of the system. In many areas in spite of favourable geo-hydrological conditions, the failure of slopes is due to the destruction of vegetation. The management and control of slope failures is necessary because of their dual role in disturbing the vegetation (see Chapter IV, Table 4.13). Firstly, in the places where slope failures occur, the loss in vegetation is an obvious feature. Apart from this loss, the spread and downward movement of loosened materials also damage the downslope vegetation. Slope failures can be checked considerably by controlling the haphazard construction of road networks which is being done without considering the local environmental conditions, and by planting vegetation in affected areas. The construction should be avoided on steep slopes and if the constructional works are necessary they should be along the lower terrace of rivers. The other important factor to be taken into account is the nature of rocks because weak and friable rocks are specially susceptible to slope failures. Thus, by avoiding areas with such rocks, roads can be constructed for the overall development of hill areas. Even then, the local environmental conditions should not be overlooked. Indeed, there is a requirement to develop new techniques for constructing the roads in these hills.

(3) Management of natural regeneration

Since long, natural regeneration has been the main source of sustained forest wealth. However, the forest wealth can be sustained only if the exploitation does not exceed regeneration. In recent times the concept of massive afforestation has originated because of a rapid decrease in the forest cover and consequent natural regeneration. Natural regeneration means production and successful growth of seedlings of plants under existing suitable environmental conditions without interference of man. In fact for the seeds falling naturally on the ground, there is a greater possibility of germination provided natural conditions of the environment are undisturbed. But during recent times, the situation has reversed. The whole process of natural regeneration is crying for immediate help. This is true not only for the Himalayan region but also for the plains and the southern uplands of Uttar Pradesh. Due to the paucity of natural regeneration, several species are endangered and many among them are on the verge of extinction.

There are many known and unknown causative factors affecting the process of natural regeneration. Among these the following are of great importance: Climate (insolation, humidity, temperature, light intensity, span of light receiving hours, precipitation and wind), soil (amount, depth, aeration, moisture level, nutrients and erosion), altitude, seed sensitivity, seed output and dispersal, biotic conditions, wildlife, forest fire and over-grazing. The following steps can augment the process of natural regeneration:

1. Protection of seedlings by controlling soil erosion, maintaining soil nutrition and moisture and checking overgrazing,

2. Pruning of canopy of augment light intensity,

3. Ground vegetation treatment,

4. Control over various diseases affecting seed output and seedling survival, and

5. Strict vigilance on forest fire.

ASPECTS OF CULTURAL ENVIRONMENT

The quantum of rapid anthropomorphization on the space-time vector of peopling and occupancy of various ecological compartments has thrown formidable challenge to eco-systems. The explosive spread of developmental catastrophe has compelled a reinterpretation of geo-ecological problems. Man, a most influensive factor of cultural environment, has adversely affected and wrought new measures of anthropogenic contouring of the nature. By adding tremendous pressure on forest resources, man is engaged in modifying ecological contours.

The scale of developmental speed and also the degree to which environment is exploited, must be correlated to understand 'what are the actual gains' and 'what are the actual losses'. Modern resource planners have started thinking about the developmental impacts and the degradation of environmental niches. This is indeed a good sign for whole of the contemporary civilization.

Developmental activities of the man, ostensibly for the man, have generated many more problems by destructing the natural vegetation. Thus, there is an urgent need to reduce the scale of man-made activities in forest exploitation. Under the heads of cultural environment following are of prime importance and may not be kept in isolation especially from the management approaches of forest resources.

(i) Population management: We need not answer the question why a National Population Policy is a must. The management of rapidly increasing population can help in reducing the burden on various resources. Population management is concerned basically with the demographic aspects of population and their spatial distribution. The increasing population means more land for buildings, more pressure on agriculture, more fuel, more fodder and more infrastructural development, etc. And all these additional requirements are directly or indirectly related with the forests. In short, population governs the level and rate of consumption of various forest products directly and various associated influences indirectly. In the state, the population is increasing at the rate of 19.82 per cent (1971-81) which will be 151.97 million in 2001 A.D., resulting in heavy pressure on forests. It will be difficult to keep pace with the clearing of forests by the present rate of plantation. For this purpose, it is essential to control population growth on the one hand and to plan plantation of trees on available denuded and vacant areas, on the other. This can be achieved by legal tools and through the creation of awareness maong the people with the help of mass-media. It is imperative that the Government should also come forward with necessary forestry courses from the primary school level itself to educate children regarding the importance of forests in shaping the environment and economy.

(ii) Afforestation: Afforestation is major means for restoring and enhancing the forest wealth in various covert and overt ways. Afforestation not only helps in enriching floral wealth but also in maintaining forest environment. By planting more and more trees, the increasing environmental problems can be mitigated to a great extent. For the purpose of afforestation, the following inventories are prerequisites:

(a) Area to be planted - areal coverage,
(b) Nature of rocks,
(c) Idea of slope gradient and slope,
(d) Soil status, and
(e) Climatic conditions such as rainfall, temperature, etc.

A knowledge of the above variables helps in the selection of suitable species as well as area to be planted. To achieve better results a measure of protection of the planted area is also important. The plantation should be based on priority and needs of the local people and the local environment.

Priority Parameters

I. Priority

(a) Hill areas,
(b) Southern uplands,
(c) Ravine lands in the Yamuna - Chambal river areas,
(d) River catchment areas,
(e) Village lands,
(f) Road, rail, canal sides, and
(g) Periphery of urban areas.

II. Hierarchical Priorities

(a) Hill areas

(i) already denuded areas,
(ii) areas under fast denudation,
(iii) areas from where forests have been cleared-off and
(iv) village localities in the form of orchards.

(b) Southern upland

(i) areas of active fluvial activities,
(ii) area without vegetation,
(iii) catchment slopes, and
(iv) edge of the uplands from where slopes start and flatness runs.

(c) Ravine lands in the Yamuna-Chambal ravir area

These ravine lands are unsuitable for agriculture and horticulture because of the very high cost of reclamation. Therefore, such lands could be best utilized for the development of forestry and grasslands.

(d) River catchment areas

(i) catchments having either no or poor vegetation,
(ii) catchments having large population pressure, and
(iii) catchments of higher anthropogenic activities.

(e) Village lands

(i) village lands having no forests,

(ii) village lands having poor forets, and
(iii) villarge lands affected by erosion.

(f) Road, Rail and Canal sides

(i) roads, rails and canals in and around flood affected areas,
(ii) roads, rails and canals in hill regions, and
(iii) roads, rails and canals of water-logging areas.

(g) Urban periphery

Green belts especially for those towns which have high industrial and commercial activities.

Under intensity state policy of vegetal cover the following areas can be selected for plantation:

(a) Dense plantation

(i) Dense plantation in the areas where cattle pressure is high,
(ii) Dense plantation in fragile and weak rock structure areas,
(iii) Dense plantation in the areas where soil erosion is high,
(iv) Dense plantation along river sides, and
(v) Dense plantation in the hill areas from where contour heights vary.

(b) Moderate plantation

(i) where vegetation is sparse,
(ii) where inter-plant distance is higher,
(iii) where cattle pressure is less,
(iv) where demands are meagre, and
(v) where moderate soil erosion process is active.

(c) Plantation in the areas where already forest exists:

(i) by removing old trees,
(ii) by removing diseased trees, and
(iii) by removing environmentally unsuitable species.

In certain areas plantation can be very useful to solve various problems of soil erosion and noise pollution etc. The following areas should be preferably considered for belt plantation:

(i) Plantation of trees in thick belts in the hill areas where inter-contour spacings are low,
(ii) In areas where rock failures occur usually, a belt of plants above and a belt of plants down the positions from where slope failures occur,
(iii) In the border areas of river basin having shallow soil, and
(iv) In the urban areas where highways are near the settlements.

There are several other measures of significance which should be considered in plantation planning, such as:

(i) needs of local/regional people,
(ii) short and long term industrial needs, and
(iii) economic and ecologic suitability of species composition.

Research should be conducted to find out suitable plant species for different bio-climatic regions. Care should be taken that such plant species be grown on village lands which can provide oilseeds, honey and seasonal fruits. In the materialistic society man has forgotton that there are other lives also on planet earth. We have not spared any component of environment for their survival in a healthy and prosperous condition. In forest management programmes emphasis should be given to plantation of those species which can provide leaf fodder. It is advisable to grow fast regenerating scrubs of such variety which can provide firewood on a sustained basis. There are certain plant species, which provide organic manure such as Burans (oak) in mountainous region. It will provide us a rid-off from those fertilizers which create a number of after effects such as soil pollution and entrophication of water-bodies. Lastly, to fulfil our requirements of clothes, we should develop plant species like Semal, Sahtoot, Arjun, Asan, Oak etc., which give us cotton, silk and tussore thread. Among the above mentioned plant species, some trees will also provide valuable construction woods after completing their overall productive living age.

Uttar Pradesh is quite lucky in having two great national institutes, i.e., NBRI, Lucknow and F.R.I., Dehradun, which can conduct researches in this direction and provide scientific assistance to the government. According to NBRI, canopy of a tree of Pipal discharges about 1,712 kg life saving air in one hour and obsorbs 2,252 kg., of CO_2 which is toxic to human life. Neem will maintain the humidity of the atmosphere. It has been suggested from experiments that if the trees are planted on 22 per cent of the farm area (on boundary), the practice will increase crop production by 50 per cent.

In addition, trees reduce noise pollution to a considerable extent arising from various urban activities in the crowded cities. Therefore, to control noise pollution, which is causing loss of hearing, plantation along roadsides be made on warfooting. It is also advised that trees, which provide a variety of essential goods and absorb the poisonous air of the atmosphere, must be planted wherever vacant space is available.

Table 5.3: **Some fuel, fodder and timber species suitable for U.P. hill region at various altitudes**

Sl. No.	Plant species	Vernacular name	Range of distribution in mountain (in m)
1.	Bombax malabaricum	Semal	300-1500
2.	Grewia oppositifolia	Bhimal	300-1000
3.	Aegle marmelos	Bel	upto 1000
4.	Cedrela toona	Tun	upto 1500
5.	Ziziphus jujuba	Ber	upto 800
6.	Ougenia dalbergioides	Sandan	upto 1500
7.	Butea frondosa	Dhak	upto 1200
8.	Dalbergia Sissoo	Shisham	upto 1000
9.	Cassia fistula	Amaltas	upto 1200
10.	Acacia arabica	Babul	upto 600
11.	A. catechu	Khair	upto 1200
12.	Albijia lebbek	Siris	upto 1000
13.	P. cornuta	Jamna	2000-3500
14.	Pyrus pashia	Mohl	800-2500
15.	Terminalia balerica	Bahera	upto 1200
16.	T. tomenstosa	Asin	upto 1200
17.	Adina cordifolia	Haldu	upto 1000
18.	Rhododendron arboreum	Burans	1000-3000
19.	Basia latifolia	Mahua	upto 1000
20.	Fraxinus micrantha	Angu	1600-2500
21.	Cordia oblique	Lisora	upto 1500
22.	Cinamomum tamal	Dalchini	600-2000
23.	Ficus bengalensis	Bargad	upto 1200
24.	F.religiosa	Pipal	upto 1200
25.	Juglans regia	Akhrot	1600-3000
26.	Betula utilis	Bhooj	3000-6000
27.	Salix tetrasperma	Bains	upto 1000
28.	Pinus longifolia	Chir	500-2700
29.	P. excelsa	Kail	1500-3600
30.	Cedrus deodara	Deodar	2200-2700

Source: A. N. Purohit, 'Physiological Research Inputs for Seed and Seedling Bank of Himalyana Mountain tree species for Social Forestry', Jr. of Himalayan Research and Development, Jan., 1984, pp.20-47.

The population pressure, and the rate of return approach of development through forests have resulted in the removal of tree cover, destruction of fragile land ecosystem, erosion and flooding etc. The basic requirements for survival especially in mountainous areas are disappearing at alarming rate. Table 5.3 shows a number of such species in mountains, which are growing since long and are fully adjusted with the local climatic condition. Such types of plants should only be planted in high altitude areas, as they will maintain the ecological balance in the area. However, since we have only limited information on these native species, a number of exotic species are being introduced for quick results. It has already been realized that we require to maintain not only the economic equilibrium to meet the short-term material needs of inhabitants but also the ecological equilibrium for long-term survival and stability. Therefore, selection of native multipurpose mountain tree species is very essential in Forest Management Programmes. The selected species should have adaptability to local climate and resistance to the local hazards. In addition, the species should have high seedling potential. If the above mentioned factors are considered in afforestation programmes, we will be more benefited from forests. Forest Department should provide information on seed germination and its improvement in Himalayan zone at local level to encourage/gear up afforestation programmes in high altitude area. Because of heavy deforestation in many hilly areas people have to go a long distance for getting fuel wood with considerable loss of profitable time. Therefore, in formulating any developmental strategy to strengthen their physical and socio- economic status emphasis should be given to plant suitable species at different elevations. It will not meet only their food, fodder, fuel and other domestic requirements but it will save the local environment in the long-term from massive degradation.

Management of forest will certainly beautify the hilly landscape, which will help in encouraging tourism. Trees moderate local climate, making it cooler in hot weather and warmer in cold weather. They also provide great comfort for enjoyment as well as active games which do not require large cleared areas. The living walls and leafy canopy give snugness and privacy, the forest floor a resilient carpet. Each leaf and twig and every animal adds a bit of beauty or interest. We can find stimulus or relaxation in forests in many ways which will activise efficiency to do more work in economic sector. To manage our woods, to give as much aesthetic satisfaction as possible, we need good planning and sensitive development according to scientific principles and needs.

Apart from massive afforestation in priority and need based areas, the plain districts are still deprived of valuable forest resources. Hence, it will be in the fitness of thing that social forestry programmes should be launched and implemented in such a way that it can fulfil the desired goal of poverty stricken rural masses. Considering the targets and achievements of social forestry, it has been felt that social forestry can boost the economy and maintain the ecological balance. This programme, therefore, awaits sincere efforts from the government and as well as from the public.

The government of Uttar Pradesh has given much emphasis on social forestry schemes in its plans. The U. P. Social Forestry, Project supported by the World Bank, covers 41 plain districts where there is acute shortage of forest produce, particularly required by the rural inhabitants. The main objective of this project is to improve the standard of living of the poorest class residing in villages through afforestation activities. With a total estimated cost of Rs. 400 million during the first five years the project envisages to raise about 50,000 ha of plantations, of which 16,000 ha will be along roads, 4,000 ha along rail tracks and the rest will be spread over degraded reserve forests (22,000 ha) and village Panchayat lands (8,000 ha).

Further, the special feature of this scheme will be to meet the daily demand of firewood, grass and fodder etc., of the rural population. Fodder, grass, leaf fodder, mahua flowers and fallen firewood will be allowed to be collected by the villagers. The Gaon Samaj will be given a share of at least 20 per cent income from the other products. A village committee will be constituted to take active part in management of these forests. Efforts will also be made to educate people, to protect their forests and raise their own common forests. In addition, a compaign will be simultaneously carried out to exhort the farmers to plant trees along their field boundaries and in vacant lands. In brief, these forests will be managed in such a way that villagers can develop a feeling that these forests are their own.

It is estimated that, with implementation of this project, from 5th year and onwards about 1.7 million tonnes of firewood will be available for consumption in rural parts of the state. This will in turn, save cowdung equal to nearly 8.0 million tonnes of manure to be used in the fields which will increase the production of crops. From the third year and onwards rural cottage and small-scale industries like production of tussore from Arjun plantation, silk production from leaves of Mulberry trees, cottage-scale making of paper pulp from Saijans wood, oil extraction from seeds of Kanju, Neem and Mahua, household furniture from the timber of Sissoo, Sirsa etc., will be established in a large number of rural centres. Besides the above gains to the villagers, several indirect benefits will accrue by raising village forests through social forestry programmes. It will greatly help in prevention of pollution of the atmosphere. The intensity of floods and droughts will decrease. The existing productivity of agricultural land will rise by more than 10 per cent. A habitat of birds and small wild life will be created. There will be definite increase in the underground water resources. Planting along canal banks will obviate the problem of waterlogging, which will also reduce the incidences of malaria.

In order to implement the programme there must be a sound organization which possesses adequate extension methodology. It has been observed that many useful programmes which appeared very beneficial for the villagers, here failed, not because there was anything inherently wrong in the project but because the persons who were responsible for carrying out the project did not possess adequate training for dealing with the people. Therefore, for successful completion of these programmes, it is utmost essential to manage such plans by competent personnel. Thus, it is necessary that in whichsoever state the social forestry programme is to be carried out, an independent wing within the Forest Department should be created. This has already been done in U.P. where a wing of social forestry Directorate under the Forest Department has been created under the Additional Chief Conservator of Forests with five conservators and 28 Forest Divions. It has greatly helped in launching such programmes in the state. But much more has to be done in this direction.

Thus, for managing the forest wealth properly, the observations made elsewhere can be summed up as follows:

It is imperative to assess the level and magnitude of degradation of forest cover through satellite images at regular intervals. Besides, like population census, there should also be a census for assessing regeneration of major species in different altitudinal zones/watersheds at regular intervals taking into account individuals of all ages. Simultaneously for the development and conservation of forests, there should be allocation of more funds because in all previous plans, the forest development was accorded very low priority in the hierarchy of state's and national budgets.

For proper forest development and management, it is necessary to augment the number of well trained and equipped foresters and forest guards so that forest personnel may have greater mobility and authority to check the illegal encroachments and poaching. This will facilitate implementation of the challenging programmes and policies to increase the tempo of forest management. There should also be a careful and continuous appraisal of the training programmes so that it can generate the updated trained man power at various levels with growing skill and expertise in forestry. In view of substantial amount of tangible and intangible benefits accruing from the forest, it is imperative to introduce Department of Forestry in a large number of universities. It will facilitate both the formal and non-formal teaching as well as research on forestry problems. For the development and conservation of forests, the involvement of people is necessary but not at the tail end of planning process. Therefore, awareness should be inculcated among people of every walk of life through mass-media. In our view inculcat- ing awareness and respect towards forest wealth is more important than incentives in the form of cash or kind.

To maintain the healthy and clean environment, forest should be treated as agricultural crops. In accordance, inputs like fertilizers, seed improvements, water, soil preparation and proper protection etc., should be provided. A policy for controlled grazing should be adopted and stall feeding encouraged. Proper management of degraded pasture land is also necessary. Further, the fast growing fodder species should be introduced to meet the growing demand for fodder. In addition, there should be made some viable alternatives for the livelihoods of the villagers located in and around the forest belts otherwise if their legitimate fuel, and fodder demands are not met they are bound to endanger the forest eco-systems. Alaternative strategies for local fuel should be introduced on the basis of social techno-energetic techniques so that demand of fuelwood can be minimised.

Agricultural extension practices, more especially in Himalayan zone and Southern plateau region should be looked into seriously. It is strongly suggested that at least in the Himalayan region, further extension in agricultural lands will bring a new crisis of environmental degradation. Besides, the danger arising from the replacement of oak with pine suggests that the near mono-cultural practice should be stopped and the afforestation of natural multispecies should be encouraged.

Before executing various developmental activities like construction of big dams, reservoirs, roads and bridges etc., particularly in forested areas, their overall environmental implications should be taken into account. Keeping these facts in view, it is suggested that a State Forest Ecology Commission should be set up under the Chairmanship of leading ecologists to decide the future course of development and forestry. The main task of this commission should be to find out various ecologically sensitive zones and to suggest appropriate measures for their speedy recovery.

It is hoped that if proper attention is paid the state will be richer in forest resources resulting in both economic and environmental development of the state.

Bibliography

Anonymous, **The State of India's Environment, 1982: A citizen's Report,** Centre for Science and Environment, New Delhi (1982).

Ashish, M. The Effect of Over-population and Overgrazing in the Kumaun, in J. S. Sing et al. **Science and Rural Development in Mountains,** Gyanodya Prakashan, Nainital, (1981), p.28.

Berreman, Gerald, G., Himachal: Science People and Progress, **IWCIA** Document 36, Denmark (1979) p. 28.

Bisht, H.S. et al. Organizational Structures of Rural Living Space in Kumaun Hills (India): An Ecological Perception, **Rural Systems,** Vol. II, Part (3), (1984), p. 165.

Champion, H.G. and Seth, S.K., **The Forest Types of India,** Manager of Publications Delhi (1968).

Climatological Tables of Observatories in India Government of India, Meteorological Department, Poona (1931-60).

Dabral, B.G. et.al. Some Studies on Air Temperature and Humidity inside Pinus Roxburghii and Dendrocalamus Strictus Plantation at New Forest, **Indian Forestry,** Vol. 95 (8) (1969).

Dani, A.P. et al. Landuse and Erosion Intensity in a Himalayan Watershed. An Illustration from Ramganga Catchment, **Himalayan Research and Development,** Vol. 1 (ii), pp.184-188 (1982).

Dwivedi, A.P. **Forestry in India,** Jugal Kishore and Co., Dehradun (1980).

Ehrich, C. The Next Flood Knocking at the Door: A Foreigner's View on the Ecological Crisis in India, in T. V. Singh (ed.) **Studies in Himalayan Ecology and Development Strategies.** The English Book tore, New Delhi, (1980) p.240

Forest Statistics, Uttar Pradesh, Govt. of Uttar Pradesh, Lucknow (1983).

Gupta, P.N. Afforestation Integrated Watershed Management, Torrent Control and Landuse Development Project for U.P. Himalayas and Siwaliks, **U.P. Forest Department,** (1979) pp.1-50.

Haigh, M.J. A Comparison of Sediments Accumulations Beneath Forest and Deforested Micro-Catchments. Garhwal Himalaya, **Himalayan Research and Development,** Vol.1 (11), (1982) pp.118-120.

Haigh, M. Deforestation and Disaster in Northern India, **Landuse Policy,** (1984) p.196.

Joshi, S.C. et al., **Kumaun Himalaya : A Geographical Perspec-**

tive on Resource Development, Gyanodaya Prakashan, Nainital, (1983) p. 324.

Maheswari, J.K. Some Aspects of the Plant Life of Himalaya with Reference to Conservation of Endangered Flora, in T. V. Singh (ed.), **Studies in Himalayan Ecology and Development Strategies,** The English Book Store, New Delhi (1980) pp. 112-16.

Marriot, Me Kin Technological Changein Overdeveloped Rural Areas, In L.W. Shannon (ed.), **Underdeveloped Areas.** Harper and Bros, New York, (1957) pp. 423-33

Moddie, A.D. Himalayan Environment, In J.S. Lal (ed.), **The Himalaya Aspects of Change,** Oxford University Press, New Delhi, (1981) p. 341.

Mishra, D.K. and C.C. Pant The Lurking Danger of Landslide: Nainital crying for Help, **Himalayan Research Development,** Vol. 1 (11), pp. 181-83.

Mithal, R.S. et. al., Environmental Impacts of the Ramganga Dam Project, **The National Geographical Jr. of India,** Vol. 30(2), p.85.

National Remote Sensing Agency, Hyderabad Nationwide mapping of Forest and Non-forest Areas Using Landsat False Colour Composites for the Periods 1972-75 and 1980-82 (Project Report), Vo.1 (1983).

Negi, S.S. Environmental Problems in the Himalayas, Dehradun (1982) pp. 35-36.

Nossin, J.J. Comparative Study of the Kalagarh Landslip Southern Himalaya, **Annals of Geomorphology,** Vo.II, (1967) pp. 356-367.

Pal, B.P. Ecological Balance - Programme and Policies to meet the challenge of 2000 A.D., **Souvenir,** 12th International Congress of Soil Sciences, New Delhi, India, Feb. 8-16, (1982) pp. 124-7.

Puri, G.S. **Indian Forest Ecology** (in two volumes), Oxford Book and Stationary Co., New Delhi (1960).

Purohit, A.N. Physiological Research Inputs for Seed and Seedling, Bank of Himalayan Mountain Tree Species for Social Forestry, **Journal of Himalayan Research and Development,** January issue, (1984) pp. 20/47.

Report of National Commission on Floods Vol.II, **Ministry of Energy and Irrigation,** Govt. of India, New Delhi (1980).

Singh, A.K. and Singh, J. Deforestation in Uttar Pradesh: An Environmental Threat, H.H. Singh et al. (ed.), **Geography and Environment Issues and Challenges,** Concept Publishing Company, New Delhi, (1985), p. 70.

Singh, A.K. and Rawat, D.S. Depletion of Oak Forest Threatening Springs; An Exploratory Survey, **National Geographical Jr.** of India, Vol.31 (1), (1985), pp.44-48.

Singh, J.S. and O.P. Chaturbedi (1981). Ecology, Environment and

Development, in J.S. Singh et al. (ed.). **Science and Rural Development in Mountains**, Gyanodaya Prakashan, Nainital, p.60.

Singh, O.P. (ed). **The Himalaya: Nature, Man and Culture,** Rajesh Publications, New Delhi (1983).

Singh, O.P. Physiographic Regions of Uttar Pradesh, **Prajna**, vol. 23 (part 2) and 24 (Part I), (1978), pp. 213-225.

Singh, S.P. Rural Eco-systems in Development in the Himalaya in J.S. Singh et al. (ed.), **Science and Rural Development in Mountains,** Gyanodaya Prakashan, Nainital (1981).

Singh, J.S. et al. Man and Forests: A Central Himalayan case Study, **AMBIO**, Vol. 13(2), (1984), pp.80-7.

Singh, J.S. et al. Replacement of Oak Forest with Pine in the Himalaya affects the Nitrogen Cycle; **Nature**, Vol. 311 (5981), (1984), pp. 54-56.

Statistical Diary, U.P. and State Income Estimates Bulletin No.194 (1982).

Thaper, S.D. **India's Forest Resources,** Birla Institute of Scientific Research Macmillan and Co., Ltd., New Delhi, (1975).

Tiwari, K.M. Forest Resources - Management and Conservation, **Souvenir**, 12th International Congress of Soil Sciences, New Delhi, India, Feb. 8-16, (1982), pp. 34.42.

Tiwari, A.R. Geography of **Uttar Pradesh,** National Book Trust, India New Delhi (1971).

Tripathi, V. **Geography of Uttar Pradesh** in Hindi, Kitab Ghar, Kanpur

Troup, R.S. Pines longifolia, Roxba sylviclutural study, Ind. For. Mem., 1, 1921. The Silviculture of Indian Trees, Oxford (1916).

Valdia, K.S. Accelerated Erosion and Landslide Prone Zones in the Central Himalaya Region, In J.S. Singh (ed.), **Environmental Regeneration in Himalaya,** CHEA, Gyanodaya Prakashan, Nainital, (1985), p. 12.

Vohra, C.P. Himalayan Glaciers in J.S. Lall (ed), **The Himalaya Aspects of Change,** Oxford University Press, New Delhi, (1981), p.85.

Yadav, R.S. Environmental Problems and Eco-Development Strategy in South Mirzapur, **The National Geographical Journal of India,** Vol. 30(4), (1984) p. 262.

Index